Noncommutative Curves of Genus Zero
Related to Finite Dimensional Algebras

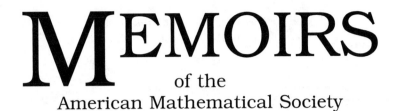

MEMOIRS
of the
American Mathematical Society

Number 942

Noncommutative Curves of Genus Zero
Related to Finite Dimensional Algebras

Dirk Kussin

September 2009 • Volume 201 • Number 942 (first of 5 numbers) • ISSN 0065-9266

American Mathematical Society
Providence, Rhode Island

2000 *Mathematics Subject Classification.*
Primary 14H45, 16G10; Secondary 14H60, 14A22.

Library of Congress Cataloging-in-Publication Data

Kussin, Dirk, 1967–
    Noncommutative curves of genus zero : related to finite dimensional algebras / Dirk Kussin.
      p. cm. — (Memoirs of the American Mathematical Society, ISSN 0065-9266 ; no. 942)
    "Volume 201, number 942 (first of 5 numbers)."
    Includes bibliographical references and index.
    ISBN 978-0-8218-4400-7 (alk. paper)
    1. Curves, Algebraic. 2. Representations of rings (Algebra) 3. Noncommutative algebras.
I. Title.

QA565.K87   2009
516.3′52—dc22
                                             2009019382

Memoirs of the American Mathematical Society

This journal is devoted entirely to research in pure and applied mathematics.

Subscription information. The 2009 subscription begins with volume 197 and consists of six mailings, each containing one or more numbers. Subscription prices for 2009 are US$709 list, US$567 institutional member. A late charge of 10% of the subscription price will be imposed on orders received from nonmembers after January 1 of the subscription year. Subscribers outside the United States and India must pay a postage surcharge of US$65; subscribers in India must pay a postage surcharge of US$95. Expedited delivery to destinations in North America US$57; elsewhere US$160. Each number may be ordered separately; *please specify number* when ordering an individual number. For prices and titles of recently released numbers, see the New Publications sections of the *Notices of the American Mathematical Society*.

Back number information. For back issues see the *AMS Catalog of Publications*.

Subscriptions and orders should be addressed to the American Mathematical Society, P. O. Box 845904, Boston, MA 02284-5904 USA. *All orders must be accompanied by payment.* Other correspondence should be addressed to 201 Charles Street, Providence, RI 02904-2294 USA.

Copying and reprinting. Individual readers of this publication, and nonprofit libraries acting for them, are permitted to make fair use of the material, such as to copy a chapter for use in teaching or research. Permission is granted to quote brief passages from this publication in reviews, provided the customary acknowledgment of the source is given.

Republication, systematic copying, or multiple reproduction of any material in this publication is permitted only under license from the American Mathematical Society. Requests for such permission should be addressed to the Acquisitions Department, American Mathematical Society, 201 Charles Street, Providence, Rhode Island 02904-2294 USA. Requests can also be made by e-mail to `reprint-permission@ams.org`.

Memoirs of the American Mathematical Society (ISSN 0065-9266) is published bimonthly (each volume consisting usually of more than one number) by the American Mathematical Society at 201 Charles Street, Providence, RI 02904-2294 USA. Periodicals postage paid at Providence, RI. Postmaster: Send address changes to Memoirs, American Mathematical Society, 201 Charles Street, Providence, RI 02904-2294 USA.

Dedicated to the memory of my beloved partner
Gordana Stanić

Contents

Abstract

In these notes we investigate noncommutative smooth projective curves of genus zero, also called exceptional curves. As a main result we show that each such curve $\mathbb{X}$ admits, up to some weighting, a projective coordinate algebra which is a not necessarily commutative *graded factorial domain* R in the sense of Chatters and Jordan. Moreover, there is a natural bijection between the points of $\mathbb{X}$ and the homogeneous prime ideals of height one in R, and these prime ideals are principal in a strong sense.

Curves of genus zero have strong applications in the representation theory of finite dimensional algebras being natural index sets for one-parameter families of indecomposable modules. They play a key role for an understanding of the notion of tameness and conjecturally for an extension of Drozd's Tame and Wild Theorem to arbitrary base fields. The function field of $\mathbb{X}$ agrees with the endomorphism ring of the unique generic module over the associated tame hereditary algebra. This skew field is of finite dimension over its centre which is an algebraic function field in one variable. As another main result we show that the function field is commutative if and only if the multiplicities determined by the homomorphism spaces from line bundles to simples sheaves (originally defined by Ringel for tame hereditary algebras) are equal to one for every point.

The study provides major insights into the nature of arithmetic complications in the representation theory of finite dimensional algebras that arise if the base field is not algebraically closed.

Received by the editor March 19, 2006 and in revised form March 18, 2007.

2000 *Mathematics Subject Classification.* Primary 14H45, 16G10; Secondary 14H60, 14A22, 16S38.

Key words and phrases. noncommutative curve, genus zero, exceptional curve, one-parameter family, separating tubular family, tame bimodule, canonical algebra, tubular algebra, orbit algebra, graded factorial domain, efficient automorphism, ghost automorphism.

Introduction

Curves of genus zero. In these notes we study noncommutative curves of genus zero. By a curve we always mean a smooth, projective curve defined over a field k. A noncommutative curve is given by a small connected k-category $\mathcal{H}$ which shares the properties with the category $\mathrm{coh}(\mathbb{X})$ of coherent sheaves over a smooth projective curve $\mathbb{X}$, listed below:

- $\mathcal{H}$ is abelian and each object in $\mathcal{H}$ is noetherian.
- All morphism and extension spaces in $\mathcal{H}$ are of finite k-dimension.
- There is an autoequivalence τ on $\mathcal{H}$ (called Auslander-Reiten translation) such that Serre duality $\mathrm{Ext}^1_{\mathcal{H}}(X,Y) = \mathrm{D}\,\mathrm{Hom}_{\mathcal{H}}(Y, \tau X)$ holds, where $\mathrm{D} = \mathrm{Hom}_k(-,k)$.
- $\mathcal{H}$ contains an object of infinite length.

It follows from Serre duality that $\mathcal{H}$ is a hereditary category, that is, $\mathrm{Ext}^n_{\mathcal{H}}$ vanishes for all $n \geq 2$. Let $\mathcal{H}_0$ be the Serre subcategory of $\mathcal{H}$ formed by the objects of finite length. Then $\mathcal{H}_0 = \coprod_{x \in \mathbb{X}} \mathcal{U}_x$ (for some index set $\mathbb{X}$) where $\mathcal{U}_x$ are connected uniserial categories, called tubes. The objects in $\mathcal{U}_x$ are called concentrated in x. Of course, any curve should also have the following property.

- $\mathbb{X}$ consists of infinitely many points.

We call $\mathbb{X}$, equipped with $\mathcal{H}$, a noncommutative (smooth, projective) curve.

It follows from the axioms (see [**74**]) that the quotient category $\mathcal{H}/\mathcal{H}_0$ is the category of finite dimensional vector spaces over some skew field $k(\mathcal{H})$, called the function field. We denote it also by $k(\mathbb{X})$. The dimension over $k(\mathcal{H})$ induces the rank of objects in $\mathcal{H}$. The full subcategory of $\mathcal{H}$ of objects which do not contain a subobject of finite length is denoted by $\mathcal{H}_+$; these objects themselves are called (vector) bundles. Bundles of rank one are called line bundles. The category $\mathcal{H}$ has the Krull-Remak-Schmidt property, that is, each object is a finite direct sum of essentially unique indecomposable objects. Moreover, each indecomposable object lies either in $\mathcal{H}_+$ or in $\mathcal{H}_0$.

In the classical case where $\mathbb{X}$ is a smooth projective curve with structure sheaf $\mathcal{O}$, the genus of $\mathbb{X}$ is zero, that is, $\dim_k \mathrm{Ext}^1_{\mathbb{X}}(\mathcal{O}, \mathcal{O}) = 0$, if and only if the category $\mathcal{H} = \mathrm{coh}(\mathbb{X})$ contains a tilting object [**69**]. This is an object $T \in \mathcal{H}$ with $\mathrm{Ext}^1_{\mathcal{H}}(T,T) = 0$ and such that $\mathrm{Hom}_{\mathcal{H}}(T,X) = 0 = \mathrm{Ext}^1_{\mathcal{H}}(T,X)$ only holds for $X = 0$.

We therefore say that a noncommutative curve $\mathcal{H}$ is of (absolute) genus zero if

- $\mathcal{H}$ contains a tilting object.

Thus, a noncommutative curve of genus zero is just an exceptional curve as defined in [**68**], a term which we will mainly use in these notes. (These curves are called "exceptional" since the existence of a tilting object is equivalent to the existence of a complete exceptional sequence of objects in $\mathcal{H}$.) In the case of genus zero the request that there are infinitely many points is automatic.

In this setting noncommutativity occurs in two different styles:

(1) The curves are allowed to be "weighted" which gives a parabolic structure on $\mathcal{H}$. This means that there are some points x in which more than one simple object is concentrated. Such a point x is called exceptional; the other points are called homogeneous. We emphasize that for the weighted curves additionally a genus in the orbifold sense (called virtual genus in [66]) is of importance.

(2) There is a another kind of noncommutativity of an arithmetic nature, determined by the function field $k(\mathcal{H})$. This skew field is commutative only in very special cases.

The first kind of noncommutativity arising by weights is well-known and the phenomenon is described in its pure form by the weighted projective lines[1] (over an algebraically closed field) defined by Geigle-Lenzing [34]. Each weighted curve of genus zero admits only finitely many exceptional points and has an underlying homogeneous curve of genus zero (where all points are homogeneous) from which it arises by so-called insertion of weights. Since this homogeneous curve has the same function field, the homogeneous case and the associated arithmetic effects of noncommutativity are the main topic of these notes.

In the following we assume this homogeneous case, which can be also expressed in the following way.

- For all simple objects $S \in \mathcal{H}$ we have $\mathrm{Ext}^1_{\mathcal{H}}(S, S) \neq 0$ (equivalently, $\tau S \simeq S$).

Such a homogeneous curve $\mathcal{H}$ has genus zero if and only if $\mathrm{Ext}^1_{\mathcal{H}}(L, L) = 0$ for one, equivalently for all line bundles L (which follows from [74]). In this case the function field $k(\mathcal{H})$ is of finite dimension over its centre which is an algebraic function field in one variable [7]. Moreover, there is a tilting object T which consists of two indecomposable summands, a line bundle L and a further indecomposable bundle $\overline{L}$ so that $\mathrm{Hom}_{\mathcal{H}}(L, \overline{L}) \neq 0$. The endomorphism ring $\mathrm{End}_{\mathcal{H}}(T)$ is a tame hereditary bimodule k-algebra. This underlying bimodule is given as $_{\mathrm{End}(\overline{L})}\mathrm{Hom}_{\mathcal{H}}(L, \overline{L})_{\mathrm{End}(L)}$.

We always consider $\mathcal{H}$ together with a fixed line bundle L which we consider as a structure sheaf. This yields a projective coordinate algebra for $\mathcal{H}$, depending on the choice of a suitable endofunctor σ on $\mathcal{H}$, and given as the orbit algebra with respect to L and σ defined as

$$\Pi(L, \sigma) = \bigoplus_{n \geq 0} \mathrm{Hom}_{\mathcal{H}}(L, \sigma^n L),$$

with multiplication given by the rule

$$g * f \overset{def}{=} \sigma^m(g) \circ f,$$

where $f \in \mathrm{Hom}(L, \sigma^m L)$ and $g \in \mathrm{Hom}(L, \sigma^n L)$. Formation of orbit algebras is a standard tool for obtaining projective coordinate algebras in algebraic geometry

---

[1]Even though in all of these cases we have graded coordinate rings and function fields which are commutative, these curves are nonetheless noncommutative since the coherent sheaves over an affine part correspond to (finitely generated) modules over a ring that is in general not commutative.

(although not under this name) and is frequently used in representation theory, see [7, 65, 49]. Note that $\Pi(L, \sigma)$ typically is noncommutative. M. Artin and J. J. Zhang used orbit algebras to define noncommutative projective schemes [2] and to prove an analogue of Serre's theorem [102].

Let for example σ be the inverse Auslander-Reiten translation τ^-. Then it is easy to see that the pair (L, τ^-) is a so-called ample pair ([2, 105]), and thus by the theorem of Artin-Zhang [2, Thm. 4.5]

$$\mathcal{H} \simeq \frac{\mathrm{mod}^{\mathbb{Z}}(\Pi(L, \tau^-))}{\mathrm{mod}_0^{\mathbb{Z}}(\Pi(L, \tau^-))},$$

the quotient category modulo the Serre subcategory of $\mathbb{Z}$-graded modules of finite length. Hence $\Pi(L, \tau^-)$ is a projective coordinate algebra for $\mathbb{X}$, and it coincides with the (small) preprojective algebra defined in [7]. However the graded algebras constructed in this way are often not practical for studying the geometry of $\mathbb{X}$ explicitly. For example, in the case of the projective line $\mathbb{X} = \mathbb{P}^1(k)$ over k (understood in the scheme sense) we have

$$\Pi(L, \tau^-) = k[X^2, XY, Y^2],$$

which consists of the polynomials in X and Y of even degree. This algebra is a projective coordinate algebras for $\mathbb{P}^1(k)$, as is the full polynomial algebra $k[X, Y]$, graded by total degree. This example illustrates the well-known fact that projective coordinate algebras are not uniquely determined, and also that some projective coordinate algebras are more useful than others. Of the two, only $k[X, Y]$ is graded factorial.

Main results. We show that there exists a graded factorial coordinate algebra in general, given as orbit algebra $\Pi(L, \sigma)$ for a suitable autoequivalence σ on $\mathcal{H}$. Of course, one has to replace the usual factoriality by a noncommutative version.

The geometry of $\mathbb{X}$ is given by the hereditary category $\mathcal{H}$. For this an understanding of the interplay between vector bundles and objects of finite length is important. In particular, with the structure sheaf L, for each point $x \in \mathbb{X}$ and the corresponding simple object $S_x \in \mathcal{U}_x$ the bimodule

$$_{\mathrm{End}(S_x)} \mathrm{Hom}(L, S_x)_{\mathrm{End}(L)}$$

is of interest. By Serre duality this is equivalent to studying the bimodule

$$_{\mathrm{End}(L)} \mathrm{Ext}^1(S_x, L)_{\mathrm{End}(S_x)},$$

and this leads directly to the universal extension

$$0 \longrightarrow L \xrightarrow{\pi_x} L(x) \longrightarrow S_x^{e(x)} \longrightarrow 0$$

with the multiplicity (originally defined by Ringel in [90])

$$e(x) = [\mathrm{Ext}^1(S_x, L) : \mathrm{End}(S_x)].$$

The above universal extension (for L) is a special case of a more general construction which leads to the tubular shift automorphism σ_x of $\mathcal{H}$, sending an object A to $A(x)$.

We realize the kernels π_x (for each $x \in \mathbb{X}$) as homogeneous elements in a suitable orbit algebra. This is accomplished by an automorphism σ on $\mathcal{H}$ which we call efficient (in 1.1.3). We show that such an automorphism always exists and has the property that for any x the middle term $L(x)$ in the universal extension is of the form $L(x) \simeq \sigma^d(L)$ for some positive integer d, depending on x.

The following theorem provides an explicit one-to-one correspondence between points of $\mathbb{X}$ and homogeneous prime ideals of height one in $\Pi(L,\sigma)$, given by forming universal extensions.

THEOREM. *Let $R = \Pi(L,\sigma)$ with σ being efficient. Let S_x be a simple sheaf concentrated in the point $x \in \mathbb{X}$. Let*

$$0 \longrightarrow L \xrightarrow{\pi_x} \sigma^d(L) \longrightarrow S_x^e \longrightarrow 0$$

be the S_x-universal extension of L. Then the element π_x is normal in R, that is, $R\pi_x = \pi_x R$. Furthermore, $P_x = R\pi_x$ is a homogeneous prime ideal of height one.

Moreover, for any homogeneous prime ideal $P \subset R$ of height one there is a unique point $x \in \mathbb{X}$ such that $P = P_x$.

In this way $\mathbb{X}$ becomes the projective prime spectrum of R. See 1.2.3 and 1.5.1 for the complete statements.

Since a *commutative* noetherian domain is factorial if and only if each prime ideal of height one is principal, we say that a noetherian graded domain R, not necessarily commutative, is a (noncommutative) graded factorial domain if each homogeneous prime ideal of height one is principal, generated by a normal element. This is a graded version of a concept introduced by Chatters and Jordan [**13**].

COROLLARY. *Each homogeneous exceptional curve admits a projective coordinate algebra which is graded factorial.*

The following results clarify the role of the multiplicities $e(x)$. The conclusion is that they measure noncommutativity ("skewness") in several senses:

THEOREM. *The function field of $\mathbb{X}$ is commutative if and only if all multiplicities are equal to one.*

See 4.3.1 for the complete statement; the commutative function fields are explicitly determined. Moreover:

- The multiplicities $e(x)$ are bounded from above by the square root $s(\mathbb{X})$ of the dimension of the function field over its centre. More precisely, if $e^*(x)$ denotes the square root of the dimension of $\mathrm{End}(S_x)$ over its centre, then always $e(x) \cdot e^*(x) \leq s(\mathbb{X})$, and equality holds for all points x except finitely many (2.2.13 and 2.3.5).
- In the graded factorial algebra R we have unique factorization in the sense that each *normal* homogeneous element is an (essentially unique) product of prime elements (which are by definition homogeneous generators of prime ideals of height one). In contrast to the commutative case, a prime element π_x may factorize into a product of several irreducible elements. The number of these factors is essentially given by $e(x)$ (see 1.6.5 and 1.6.6).
- We describe the localization R_P at a prime ideal P. It turns out that R_P is a local ring if and only if the corresponding multiplicity $e(x)$ is one; otherwise R_P is not even semiperfect (2.2.15).

Another surprising phenomenon due to noncommutativity is the occurrence of so-called ghost automorphisms. Denote by $\mathrm{Aut}(\mathbb{X})$ the group of all (isomorphism classes of) automorphisms of the category $\mathcal{H}$ fixing the structure sheaf L. Let $R = \Pi(L,\sigma)$ be the orbit algebra formed with respect to an efficient automorphism σ. Every prime element $\pi_y \in R$ (that is, a normal element generating the prime ideal P_y associated to the point y) induces a graded algebra automorphism γ_y on R, given by the formula $r\pi_y = \pi_y \gamma_y(r)$. This in turn induces an automorphism $\gamma_y^* \in \mathrm{Aut}(\mathbb{X})$ whose action on the set of all points of $\mathbb{X}$ is invisible, but it is a non-trivial element of $\mathrm{Aut}(\mathbb{X})$ if (under an additional assumption, see 3.2.4) for all units u the element $\pi_y u$ is not central. This means that the functor γ_y^* fixes all objects but acts non-trivially on morphisms. Such a functor we call a ghost automorphism.

The simplest example in which this effect arises is given by the curve $\mathbb{X}$ with underlying bimodule $M = {}_\mathbb{C}(\mathbb{C} \oplus \overline{\mathbb{C}})_\mathbb{C}$ over $k = \mathbb{R}$, where $\mathbb{C}$ acts from the right on the second component via conjugation. A projective coordinate algebra is given by the graded twisted polynomial ring $R = \mathbb{C}[X; Y, \bar{\cdot}\,]$, graded by total degree, where X is a central variable and for the variable Y we have $Ya = \bar{a}Y$ for all $a \in \mathbb{C}$. We write $R = \mathbb{C}[X, \overline{Y}]$. Then Y is a prime element which is not central (up to units). It follows that complex conjugation induces a ghost automorphism of $\mathbb{X}$. Moreover, denote by σ_x and σ_y the (efficient) tubular shifts corresponding to the points x and y associated with the prime ideals generated by X and Y, respectively. Then $\mathbb{C}[X, \overline{Y}] = \Pi(L, \sigma_x)$ holds.

The following theorem expresses the interrelation between various automorphisms in more detail.

THEOREM. *Let* $R = \Pi(L, \sigma)$, *where* σ *is efficient. Let* π_y *be a prime element of degree* d *in* R, *associated to the point* y *and* γ_y *the induced graded algebra automorphism. Let* σ_y *be the tubular shift associated to* y. *Then there is an isomorphism of functors* $\sigma_y \simeq \sigma^d \circ \gamma_y^*$.

The theorem contains important information about the structure of the Picard group $\mathrm{Pic}(\mathbb{X})$, defined as the subgroup of $\mathrm{Aut}(\mathcal{H})$ generated by all tubular shifts σ_x ($x \in \mathbb{X}$). In particular, in contrast to the algebraically closed case, the Picard group may not be isomorphic to $\mathbb{Z}$.

In Chapter 5 we develop a technique which allows explicit calculation of the automorphism group $\mathrm{Aut}(\mathbb{X})$ in many cases. We illustrate this for the preceding example, where $R = \mathbb{C}[X, \overline{Y}]$. The ghost group is the subgroup of $\mathrm{Aut}(\mathbb{X})$ consisting of all ghost automorphisms.

PROPOSITION. *Let* $\mathbb{X}$ *be the homogeneous curve with projective coordinate algebra* $R = \mathbb{C}[X, \overline{Y}]$. *Then* $R = \Pi(L, \sigma_x)$, *and* $\mathrm{Aut}(\mathbb{X})$ *is generated by*

- *the automorphism* γ_y^* *of order two, induced by complex conjugation, generating the ghost group;*
- *transformations of the form* $Y \mapsto aY$ *for* $a \in \mathbb{R}_+$;
- *the automorphism induced by exchanging* X *and* Y.

Moreover, the Picard group $\mathrm{Pic}(\mathbb{X})$ *is isomorphic to* $\mathbb{Z} \times \mathbb{Z}_2$, *and for the Auslander-Reiten translation the following formula holds true:*

$$\tau = \sigma_x^{-1} \circ \sigma_y^{-1} = \sigma_x^{-2} \circ \gamma_y^*.$$

See Sections 5.3 and 5.4 for more general statements. In general the functorial properties of the Auslander-Reiten translation have not been extensively studied. The preceding result shows that interesting effects appear. On objects the Auslander-Reiten translation τ acts like σ_x^{-2}, which agrees with the degree shift by -2. But on morphisms the ghost automorphism induced by complex conjugation enters the game.

So far in this introduction we have concentrated on the homogeneous case. These notes also deal with the weighted case. The following results show that the problem of determining the geometry of an exceptional curve can often be reduced to the homogeneous case.

- We show that insertion of weights into a *central* prime element in a graded factorial coordinate algebra preserves the graded factoriality; the resulting graded algebra is a projective coordinate algebra of a (weighted) exceptional curve (6.2.4).
- The automorphism group of a (weighted) exceptional curve is given by the automorphisms of the underlying homogeneous curve preserving the weights (6.3.1). In particular, both curves have the same ghost group.

The insertion of weights is particularly important for our treatment of the tubular case in Chapter 8. The tubular exceptional curves have a strong relationship to elliptic curves. They are defined by the condition that the so-called virtual (orbifold) genus is one. The main feature of the tubular case is that, very similar to Atiyah's classification of vector bundles over an elliptic curve, $\mathcal{H}$ consists entirely of tubular families. In fact, there is a linear form deg, called the degree, which together with the rank rk defines the slope $\mu(X) = \frac{\deg X}{\operatorname{rk} X}$ of (non-zero) objects X in $\mathcal{H}$. Denote for $q \in \widehat{\mathbb{Q}} = \mathbb{Q} \cup \{\infty\}$ by $\mathcal{H}^{(q)}$ the additive closure of indecomposable objects in $\mathcal{H}$ of slope q. Then $\mathcal{H}$ is the additive closure of all $\mathcal{H}^{(q)}$, where $q \in \widehat{\mathbb{Q}}$.

In case the base field is algebraically closed all the tubular families $\mathcal{H}^{(q)}$ are isomorphic to each other as categories, and moreover each is parametrized by the curve $\mathbb{X}$. The reason for this is that in this case the natural action of the automorphism group $\operatorname{Aut}(D^b(\mathcal{H}))$ on the set $\widehat{\mathbb{Q}}$ is transitive. This is not true in general over an arbitrary base field. We show in Chapter 8 that in general this action may have up to three orbits [**53, 59**]. Accordingly, there are up to three different tubular exceptional curves which are Fourier-Mukai partners.

Another interesting effect treated in the same chapter is the occurrence of line bundles which are not exceptional. Over an algebraically closed field each line bundle L over an exceptional curve $\mathbb{X}$ is exceptional, that is, satisfies $\operatorname{Ext}^1(L, L) = 0$. But this does not extend to arbitrary base fields, the simplest counterexamples existing in the tubular case. We characterize the tubular cases where non-exceptional line bundles exist and show how they can be determined explicitly (Section 8.5).

Applications to finite dimensional algebras. The study of noncommutative curves of genus zero has strong applications in the representation theory of finite dimensional algebras. Conjecturally these curves yield the natural parametrizing sets for one-parameter families of indecomposable modules over finite dimensional tame algebras. This is reflected by the definition of tame algebras over an algebraically closed field k, using as parametrizing curves (affine subsets of) the projective line $\mathbb{P}^1(k)$, and in a certain sense this "explains" that in Drozd's Tame and Wild Theorem [**32, 17**] only rational one-parameter families occur. Note that in the algebraically closed case $\mathbb{P}^1(k)$ is the only homogeneous curve of genus zero.

For the class of tame hereditary algebras and the class of tame canonical algebras [**92**] over an arbitrary field it is well-known that the parametrizing sets are precisely the (affine) curves of genus zero. For a tame algebra, in general more than one exceptional curve is needed to parametrize the indecomposables: there is a tubular (canonical) algebra which requires three such curves (Section 8.3).

It is important to study representation theory over arbitrary base fields since many applications deal with algebras defined over fields which are not algebraically closed. For example, the base field of real numbers is of interest for applications in analysis, the field of rational numbers for number theory, finite fields for the relationships to quantum groups (like Ringel's Hall algebra approach), etc.

When attempting to generalize statements first proven over algebraically closed fields to arbitrary base fields, three typical scenarios of different nature can be observed. Frequently statements and proofs carry over to the more general situation without essential change. Also often the statements remain true but require new proofs, frequently leading to better insights and streamlined arguments even for the algebraically closed case[2]. On the other hand, in a significant number of cases completely new and unexpected effects occur, causing the statements to fail in the general case. The present notes focus in particular on these kinds of new effects.

The representation theoretical analogues of the exceptional curves $\mathbb{X}$ and their hereditary categories $\mathcal{H}$ are given by the concealed canonical algebras [70] and their module categories $\mathrm{mod}(\Lambda)$. The link between the two concepts is given by an equivalence $\mathrm{D}^b(\mathcal{H}) \simeq \mathrm{D}^b(\mathrm{mod}(\Lambda))$ of derived categories which leads to a translation between geometric and representation theoretic notions. We illustrate this in the typical case where Λ is a tame hereditary algebra: the subcategory $\mathcal{H}_0$ of objects of finite length corresponds to the full subcategory $\mathcal{R}$ of $\mathrm{mod}(\Lambda)$ formed by the regular representations. Simple objects S_x in $\mathcal{H}$ correspond to simple regular representations. Vector bundles correspond to preprojective (or preinjective) modules, line bundles L to preprojective modules P (or preinjective modules) of defect -1 (or 1, respectively). In particular, the multiplicities $e(x)$ are also definable in terms of preprojective modules of defect -1 and simple regular representations. The function field of $\mathbb{X}$ agrees with the endomorphism ring of the unique generic [19] Λ-module. The importance of the generic module for the representation theory of tame hereditary algebras is demonstrated in [90]. Our results on exceptional curves all have direct applications to representation theory. In particular:

- Let Λ be a tame hereditary algebra. The (small) preprojective algebra

$$\bigoplus_{n \geq 0} \mathrm{Hom}_\Lambda(P, \tau^{-n}P),$$

  where P is a projective module of defect -1 and τ^- is the (inverse) Auslander-Reiten translation on $\mathrm{mod}(\Lambda)$, is a graded factorial domain if the underlying tame bimodule is of dimension type $(1,4)$ (or $(4,1)$)[3]. Note that the (small) preprojective algebra contains the full information on Λ and its representation theory.
- In general there are automorphisms of $\mathrm{D}^b(\mathrm{mod}(\Lambda))$ fixing all objects but acting non-trivially on morphisms, contrary to the algebraically closed case.
- A tubular algebra requires up to three different projective curves of genus zero to parametrize the indecomposable modules.

---

[2]Some examples for this can be seen in the results of Happel and Reiten about the characterization of hereditary abelian categories with tilting object ([39], generalizing [38]) and in the proof of the transitivity of the braid group action on complete exceptional sequences for hereditary Artin algebras by Ringel ([94], generalizing [20]) and by Meltzer and the author for exceptional curves ([60], summarized in Section 7.1), generalizing [78].

[3]This is also true for many tame bimodules of dimension type $(2,2)$.

- A tubular algebra admits generic modules with up to three different (non-isomorphic) endomorphism rings.
- The endomorphism ring of the generic module over a tame hereditary algebra is commutative if and only if all multiplicities are equal to one, a condition automatically satisfied over algebraically closed base fields. It is surprising that this condition, which essentially says that the morphisms between preprojective and regular representations behave "well", yields the commutativity of the generic module's endomorphism ring, and conversely.

The results on the function field also provide an explanation of the strange fact (pointed out in [90]) that a bimodule like $_\mathbb{R}\mathbb{H}_\mathbb{H}$, given by noncommutative data, leads to a commutative function field

$$\mathrm{Quot}\big(\mathbb{R}[U, V]/(U^2 + V^2 + 1)\big),$$

whereas a bimodule like $_\mathbb{Q}\mathbb{Q}(\sqrt{2}, \sqrt{3})_{\mathbb{Q}(\sqrt{2}, \sqrt{3})}$, given by commutative data, leads to a noncommutative function field, the quotient division ring of

$$\mathbb{Q}\langle U, V\rangle/(UV + VU, V^2 + 2U^2 - 3).$$

There are a number of inspiring papers dealing with tame hereditary algebras. For example, those by Dlab and Ringel on bimodules and hereditary algebras [24, 89, 27, 26, 29] (see additionally [28, 22, 23]), in particular Ringel's Rome proceedings paper [90], as well as those by Lenzing [64], Baer, Geigle and Lenzing [7], and by Crawley-Boevey [18], dealing with the structure of the parameter curves for tame hereditary algebras over arbitrary fields.

By perpendicular calculus and insertion of weights many problems for concealed canonical algebras (and in particular for tame hereditary algebras) can be reduced to the special class of tame bimodule algebras. This means that we often may restrict our attention to a tame hereditary k-algebra of the form $\Lambda = \begin{pmatrix} G & 0 \\ M & F \end{pmatrix}$, where $M = {}_FM_G$ is a tame bimodule over k, that is, the product of the dimensions of M over the skew fields F and G, respectively, equals 4. These are the analogues of the Kronecker algebra $\begin{pmatrix} k & 0 \\ k^2 & k \end{pmatrix}$, which is isomorphic to the path algebra of the following quiver.

$$\bullet \underset{\longrightarrow}{\overset{\longrightarrow}{}} \bullet$$

In this homogeneous case $\mathbb{X}$ parametrizes the simple regular representations of Λ. This situation was studied in the cited papers by Dlab and Ringel, by Baer, Geigle and Lenzing, and by Crawley-Boevey. Over the real numbers the structure of $\mathbb{X}$ as topological space is described explicitly in [24, 25, 26]. In [89, 29] and more generally in [18] an affine part of $\mathbb{X}$ is described by the simple modules over a (not necessarily commutative) principal ideal domain. In [18] additionally a (commutative) projective curve is constructed, which parametrizes the points of $\mathbb{X}$ and is the centre of the noncommutative projective curves considered in [64] and [7]. A model-theoretic approach using the Ziegler spectrum is described by Prest [86] and Krause [51, Chapter 14]. One advantage provided by the present notes is that the geometry of $\mathbb{X}$ is described in terms of graded factorial coordinate algebras. This is useful in particular for studying the properties of the sheaf category $\mathcal{H}$ by forming natural localizations (Chapter 2) and for analyzing the automorphism group of $\mathrm{D}^b(\mathcal{H})$ (Chapter 3). It is also exploited in our proof of the characterization of the commutativity of the function field in terms of the multiplicities (Section 4.3).

We have seen that several new and surprising phenomena occur when an arbitrary base field is allowed. Along the way, we will point out several interesting open problems. The following are particularly worth mentioning:

- Find graded factorial projective coordinate algebras for all weighted cases (by a suitable method of inserting weights also into non-central prime elements).
- Determine the ghost group in general. Describe the action of the Auslander-Reiten translation on morphisms in general.
- The function field $k(\mathbb{X})$ is always of finite dimension over its centre. Is the square root of this dimension always the maximum of the multiplicity function e? Describe each multiplicity $e(x)$ in terms of the function field.
- Is it true that the completions $\widehat{R}$ of the described graded factorial algebras R are factorial again?

These notes are based on the author's Habilitationsschrift with the title "Aspects of hereditary representation theory over non-algebraically closed fields" accepted by the University of Paderborn in 2004. The present version includes further recent results, in particular those concerning the multiplicities in Chapter 2.

We assume that the reader is familiar with the language of representation theory of finite dimensional algebras. We refer to the books of Assem, Simson and Skowroński [3], of Auslander, Reiten and Smalø [5], and of Ringel [91].

Acknowledgements. It is a pleasure to express my gratitude to several people. First I wish to thank Helmut Lenzing for many inspiring discussions on the subject. Most of what I know about representation theory and weighted projective lines I learned from him. He has always encouraged my interest in factoriality questions in this context which started with my doctoral thesis and the generalization [55] of a theorem of S. Mori [83]. For various helpful discussions and comments I would like to thank Bill Crawley-Boevey, Idun Reiten and Claus Ringel. In particular, some of the results concerning multiplicities were inspired by questions and comments of Bill Crawley-Boevey and Claus Ringel. I also would like to thank Hagen Meltzer for many discussions on several aspects of weighted projective lines and exceptional curves. The section on the transitivity of the braid group action is a short report on a joint work with him. I got the main idea for the definition of an efficient automorphism and thus for the verification of the graded factoriality in full generality during a visit at the Mathematical Institute of the UNAM in Mexico City when preparing a series of talks on the subject. I thank the colleagues of the representation theory group there, in particular Michael Barot and José Antonio de la Peña, for their hospitality and for providing a stimulating working atmosphere. For their useful advices and comments on various parts and versions of the manuscript I thank Axel Boldt, Andrew Hubery and Henning Krause. For her love and her patience I wish to thank my dear partner Gordana.

CHAPTER 0

Background

In this preliminary chapter we describe the setting and present the background material from the literature which will be used later. The main parts of this work will start with Chapter 1. We recommend to browse through this chapter or even start reading the work with Chapter 1 and look up items here when necessary.

0.1. Notation

We work over an arbitrary field k. If not otherwise specified, all categories will be k-categories and all functors will be k-functors and covariant. If the isomorphism classes of objects in a category $\mathcal{C}$ form a set, then we call $\mathcal{C}$ small. (This is often called skeletally-small in the literature.) If X is an object in $\mathcal{C}$ we write $X \in \mathcal{C}$ instead of $X \in \mathrm{Ob}(\mathcal{C})$.

All rings and algebras are associative with identity. If not otherwise specified, by modules we mean right modules, and all modules are unitary. The category of all R-modules is denoted by $\mathrm{Mod}(R)$. The full subcategory of finitely presented R-modules is denoted by $\mathrm{mod}(R)$. Since we will only consider noetherian situations, these are just the finitely generated modules. If R is an algebra graded by an abelian group H we denote by $\mathrm{Mod}^H(R)$ the category of H-graded R-modules; the morphisms are those of degree zero. The subcategory $\mathrm{mod}^H(R)$ is similarly defined like in the ungraded situation.

0.2. One-parameter families, generic modules and tameness

In this section we briefly recall the notions of one-parameter families and tameness. Although we will not explicitly use these facts later in the text, they serve as one of the main motivations.

In the representation theory of finite dimensional algebras certain modules often form sets with geometric structure. By the Tame and Wild Theorem of Drozd [**32**] (see also [**17**]) the indecomposable modules over a non-wild (= tame) finite dimensional algebra over an algebraically closed field $\overline{k}$ essentially lie in rational one-parameter families, that is, families indexed by (an affine part of) the projective line $\mathbb{P}^1(\overline{k})$. (We use the rather unusual notation $\overline{k}$ in order to stress that temporarily the field is assumed to be algebraically closed.)

0.2.1. Let A be a finite dimensional algebra over an algebraically closed field $\overline{k}$. Let M be a $\overline{k}[T]$-A-bimodule which is free of finite rank as left $\overline{k}[T]$-module. Consider the associated functor

$$F_M = - \otimes_{\overline{k}[T]} M : \mathrm{mod}(\overline{k}[T]) \longrightarrow \mathrm{mod}(A).$$

11

For each $\lambda \in \overline{k}$ let S_λ be the simple $\overline{k}[T]$-module $\overline{k}[T]/(T - \lambda)$. If all the images $F_M(S_\lambda)$ are indecomposable and pairwise non-isomorphic, then $\{F_M(S_\lambda)\}_{\lambda \in \overline{k}}$ is called an affine one-parameter family (of indecomposable modules).

0.2.2 (Tame algebras). Let A be a finite dimensional algebra over an algebraically closed field $\overline{k}$. Then A is called *tame*, if for each natural number d almost all indecomposable A-modules of dimension d lie in a finite number of affine one-parameter families, that is, given d there are finitely many $\overline{k}[T]$-A-bimodules M_i, free of finite rank over $\overline{k}[T]$, such that all but finitely many indecomposable A-modules of dimension d are isomorphic to $F_{M_i}(S_\lambda)$ for some i and some $\lambda \in \overline{k}$.

0.2.3 (Generic modules). In the study of one-parameter families the concept of a generic module is important ([**19**], also [**50**]). An A-module M is called *generic* [**19**], if it is indecomposable, of infinite length over A, but of finite length over its endomorphism ring. Note that for each affine one-parameter family, given by a functor F_M, a generic A-module is given by $F_M(\overline{k}(T))$, where $\overline{k}(T)$ is the field of rational functions in one variable.

Crawley-Boevey [**19**] has shown that, over an algebraically closed field, A is tame if and only if for any natural number d there is only a finite number of generic modules of endolength d. (In the latter case one also says that A is *generically tame*. This notion makes sense over any field.) He showed that in this case the generic modules correspond to the one-parameter families.

0.2.4 (The Kronecker algebra). The Kronecker algebra Λ over an algebraically closed field $\overline{k}$ provides the prototype of a tame algebra as well as of a one-parameter family. It is defined to be the path algebra of the quiver

$$\bullet \Longrightarrow \bullet$$

and is isomorphic to $\Lambda = \begin{pmatrix} \overline{k} & 0 \\ \overline{k}^2 & \overline{k} \end{pmatrix}$, where $\overline{k}^2 = \overline{k} \oplus \overline{k}$ is considered as $\overline{k}$-$\overline{k}$-bimodule.

The module category $\mathrm{mod}(\Lambda)$, as well as its Auslander-Reiten quiver, has a particular simple shape, it is trisected

$$\mathrm{mod}(\Lambda) = \mathcal{P} \vee \mathcal{R} \vee \mathcal{Q},$$

where $\mathcal{P}$ is the preprojective component, consisting of the Auslander-Reiten orbits of two projective indecomposables, $\mathcal{Q}$ is the preinjective component, consisting of the Auslander-Reiten orbits of two injective indecomposables, and $\mathcal{R}$ consists of the regular indecomposable modules, all lying in homogeneous tubes. One can say that $\mathcal{P}$ and $\mathcal{Q}$ form the discrete part of $\mathrm{mod}(\Lambda)$ and $\mathcal{R}$ forms the continuous part, since the tubes are parametrized by the projective line $\mathbb{P}^1(\overline{k})$. Moreover, if one forms the category

$$\mathcal{H} \overset{def}{=} \mathcal{Q}[-1] \vee \mathcal{P} \vee \mathcal{R}$$

inside the bounded derived category of $\mathrm{mod}(\Lambda)$, then $\mathcal{H}$ is equivalent to $\mathrm{coh}(\mathbb{P}^1(\overline{k}))$, the category of coherent sheaves over $\mathbb{P}^1(\overline{k})$.

The regular indecomposable modules of a fixed dimension form the one-parameter families for Λ (leave out one tube for an affine family). The regular part $\mathcal{R}$ itself forms a separating tubular family.

There is (up to isomorphism) precisely one generic Λ-module, given by the representation

$$\overline{k}(T) \xrightarrow[\cdot T]{1} \overline{k}(T)\,,$$

where $\overline{k}(T)$ is the field of rational functions in one variable, which is the function field of $\mathbb{P}^1(\overline{k})$; the endomorphism ring of this generic module is $\overline{k}(T)$.

Let A be a tame $\overline{k}$-algebra. By Drozd's theorem all one-parameter families for A are rational. In all known examples these parametrizations can be realized by a functor $\mathrm{mod}(\Lambda) \longrightarrow \mathrm{mod}(A)$.

0.2.5. Over an arbitrary field k there is still no convenient definition of tameness. The definition of generically tameness makes sense over any field and has many advantages, but it does not capture the geometric flavour of one-parameter families. One should expect that an extension of Drozd's Tame and Wild Theorem over arbitrary field k holds in the sense that, roughly speaking, the indecomposable finite dimensional modules over any non-wild finite dimensional k-algebra lie essentially in one-parameter families which are indexed by (affine parts of) the noncommutative curves of genus zero. The projective line is related to the Kronecker algebra, just as the noncommutative curves of genus zero are related (up to weights) to the tame bimodules $M = {}_FM_G$ and their associated hereditary algebras

$$\Lambda = \begin{pmatrix} G & 0 \\ M & F \end{pmatrix}\,,$$

which were studied by Dlab and Ringel in several papers (for example [24, 89, 29], to name a few). Therefore the tame bimodules are of fundamental importance in the study of one-parameter families. Note that in general different one-parameter families of genus zero for a fixed finite dimensional k-algebra may be induced by different tame bimodules over k, as the discussion in Chapter 8 shows.

0.2.6 (The weighted case). In general one has do deal with the so-called weighted case which leads to the study of the canonical algebras and to the weighted projective lines (as Ringel pointed out in his survey [93]). Over algebraically closed fields, the canonical algebras were defined by Ringel [91] and the weighted projective lines by Geigle and Lenzing [34]. Both definitions were later extended to arbitrary fields. In the case of the canonical algebras this was done by Ringel and Crawley-Boevey [92], in the case of the weighted projective lines by Lenzing [68] who called the more general objects exceptional curves. The canonical algebras can be characterized (essentially up to tilting equivalence) as the class of finite dimensional algebras admitting a separating tubular family [70]. These tubular families are parametrized by the exceptional curves. The tame bimodule algebras correspond to the subclass of finite dimensional algebras whose tubes are all homogeneous. So we call the tame bimodule case also the homogeneous or unweighted case, the general case also the weighted case.

By some general techniques (perpendicular calculus [35]), insertion of weights [68]) the general, weighted case can be reduced essentially to the homogeneous case. Therefore, main parts of this article are concerned with the homogeneous case.

0.3. Canonical algebras and exceptional curves

In this section we describe briefly the general class of finite dimensional algebras admitting a separating tubular family. This is the class of concealed canonical algebras, which contains the class of canonical algebras and the class of tame hereditary algebras, in particular tame bimodule algebras. These algebras have as geometric counterpart the exceptional curves. These curves correspond to the concealed canonical algebras via tilting theory, and accordingly are derived equivalent to the corresponding algebra. Thus the study of (concealed) canonical algebras is essentially equivalent to the study of exceptional curves. Since we are interested in the geometrical aspects of algebras, we prefer in this paper the usage of the language and theory of the exceptional curves.

0.3.1 (Concealed canonical algebras). Let k be a field and Σ a finite dimensional k-algebra, which is assumed to be connected. Denote by $\mathrm{mod}(\Sigma)$ the category of finitely generated right Σ-modules. Then Σ is *concealed canonical* ([**70**], see also [**104**]) if and only if $\mathrm{mod}(\Sigma)$ contains a *sincere separating exact subcategory* $\mathrm{mod}_0(\Sigma)$. This means

- Exactness. $\mathrm{mod}_0(\Sigma)$ is an exact abelian subcategory of $\mathrm{mod}(\Sigma)$, which is stable under Auslander-Reiten translation $\tau = \mathrm{D}\,\mathrm{Tr}$ and $\tau^- = \mathrm{Tr}\,\mathrm{D}$
- Separation. Each indecomposable from $\mathrm{mod}(\Sigma)$ belongs either to $\mathrm{mod}_0(\Sigma)$ or to $\mathrm{mod}_+(\Sigma)$, which consists of all $M \in \mathrm{mod}(\Sigma)$ such that $\mathrm{Hom}(\mathrm{mod}_0(\Sigma), M) = 0$, or to $\mathrm{mod}_-(\Sigma)$, which consists of all $N \in \mathrm{mod}(\Sigma)$ such that $\mathrm{Hom}(N, \mathrm{mod}_0(\Sigma)) = 0$.
- Sincerity. For each non-zero $M \in \mathrm{mod}_+(\Sigma)$ there is a non-zero morphism from M to $\mathrm{mod}_0(\Sigma)$ and for each non-zero $N \in \mathrm{mod}_-(\Sigma)$ there is non-zero morphism from $\mathrm{mod}_0(\Sigma)$ to N.
- Stability. Each projective module belongs to $\mathrm{mod}_+(\Sigma)$ and each injective module to $\mathrm{mod}_-(\Sigma)$.

0.3.2. The most prominent classes of examples are the following:

(1) The canonical algebras, as defined by Ringel and Crawley-Boevey in [**92**]. Actually, every concealed canonical algebra is tilting equivalent to a canonical algebra. A canonical algebra is defined to be the tensor algebra of a species

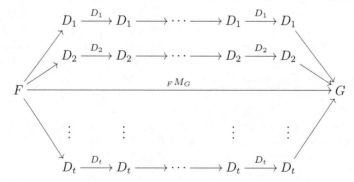

modulo certain relations (for details we refer to [**92**]). Here, $_FM_G$ is a tame bimodule (see 0.3.16 below), and there are t arms, the i-th arm of length $p_i \geq 1$, and the D_i are finite dimensional skew fields over k, with k lying in their centres. Moreover, there are F-D_i-bimodule U_i and D_i-G-bimodules V_i (k acting centrally) on the arrows starting in the source and ending in the sink, respectively.

(2) The tame hereditary algebras. In particular, the tame bimodule algebras (see 0.3.16 and 0.5.1). Actually, by the so-called insertion of weights [**68**], and by the perpendicular calculus [**35**], two processes which are inverse to each other, many problems for concealed canonical algebras can be reduced to the special class of tame bimodule algebras. (We will explain this in 0.3.16.)

0.3.3 (Separating tubular family). A sincere separating exact subcategory $\mathrm{mod}_0(\Sigma)$ defines a separating tubular family of stable tubes [**92**]: there is the coproduct of categories

$$\mathrm{mod}_0(\Sigma) = \coprod_{x \in \mathbb{X}} \mathcal{U}_x,$$

where $\mathcal{U}_x$ are connected, uniserial length categories, containing neither non-zero projective nor non-zero injective modules. The full subcategory $\mathcal{T}_x = \mathrm{ind}(\mathcal{U}_x)$ of indecomposable objects in $\mathcal{U}_x$ is called a *stable tube*. Moreover, each non-zero morphism from an object in $\mathrm{mod}_+(\Sigma)$ to an object in $\mathrm{mod}_-(\Sigma)$ factorizes through any prescribed tube $\mathcal{U}_x$.

0.3.4 (Associated hereditary category). In the preceding coproduct, $\mathbb{X}$ is an index set, which is equipped with geometric structure. In [**70**] there is defined an associated hereditary abelian k-category $\mathcal{H}$. Hereditary means that $\mathrm{Ext}^i_{\mathcal{H}}(-,-) = 0$ for all $i \geq 2$. Roughly speaking, to construct $\mathcal{H}$ one takes the union of $\mathrm{mod}_0(\Sigma)$ and $\mathrm{mod}_+(\Sigma)$ and forms inside the bounded derived category $\mathrm{D}^b(\mathrm{mod}(\Sigma))$ (see [**37**]) the closure of this union under all inverse shift automorphisms defined to tubes in $\mathrm{mod}_0(\Sigma)$. By the construction it is immediate that the categories $\mathcal{H}$ and $\mathrm{mod}(\Sigma)$ are derived equivalent.

In the most important special case when Σ is a tame bimodule algebra we describe the category $\mathcal{H}$ more explicitly in 0.5.1.

0.3.5 (Bundles/objects of finite length). Denote by $\mathcal{H}_0$ ($\mathcal{H}_+$, respectively) the full subcategory of $\mathcal{H}$ of objects of finite length (of objects, not containing objects $\neq 0$ of finite length, respectively). Then each indecomposable object in $\mathcal{H}$ is either in $\mathcal{H}_0$ or in $\mathcal{H}_+$, and $\mathrm{Hom}_{\mathcal{H}}(\mathcal{H}_0, \mathcal{H}_+) = 0$. The objects of $\mathcal{H}$ ($\mathcal{H}_+$, respectively) are also called *sheaves* (*vector bundles* or *torsionfree*, respectively). By construction of $\mathcal{H}$ we have $\mathcal{H}_0 = \mathrm{mod}_0(\Sigma) = \coprod_{x \in \mathbb{X}} \mathcal{U}_x$.

0.3.6 (Exceptional curves). $\mathbb{X}$, together with the category $\mathcal{H}$, is called an *exceptional curve* [**68**], and one sometimes writes $\mathcal{H} = \mathrm{coh}(\mathbb{X})$. This class of categories $\mathcal{H}$ is characterized independently of the construction above by the following properties:

- $\mathcal{H}$ is a connected small abelian k-category with finite dimensional morphism and extension spaces.
- $\mathcal{H}$ is hereditary and noetherian and contains no non-zero projective object.
- $\mathcal{H}$ admits a tilting object (see the following number).

0.3.7 (Tilting object). $T \in \mathcal{H}$ is called a *tilting object*, if

- $\mathrm{Ext}^1_{\mathcal{H}}(T, T) = 0$, and
- If $X \in \mathcal{H}$, then $\mathrm{Hom}_{\mathcal{H}}(T, X) = 0 = \mathrm{Ext}^1_{\mathcal{H}}(T, X)$ implies $X = 0$.

A tilting object lying in $\mathcal{H}_+$ is called a *tilting bundle*.

There exists even a tilting bundle T such that $\mathrm{End}_{\mathcal{H}}(T)$ is a canonical algebra ([**70**, Prop. 5.5]).

0.3.8 (Exceptional object). An object E in $\mathcal{H}$ is called *exceptional* if it is indecomposable and $\mathrm{Ext}^1_{\mathcal{H}}(E, E) = 0$. It follows then by an argument by Happel and Ringel [41] that $\mathrm{End}_{\mathcal{H}}(E)$ is a skew field.

0.3.9 (Serre duality). For an exceptional curve there is an autoequivalence τ on $\mathcal{H}$ such that Serre duality

$$\mathrm{Ext}^1_{\mathcal{H}}(X, Y) \simeq \mathrm{D}\,\mathrm{Hom}_{\mathcal{H}}(Y, \tau X)$$

holds functorially in $X, Y \in \mathcal{H}$, where D is the duality $\mathrm{Hom}_k(-, k)$.

Since the category $\mathcal{H}$ is hereditary, the (bounded) derived category $\mathrm{D}^b(\mathbb{X}) \overset{def}{=} \mathrm{D}^b(\mathcal{H}) = \mathrm{D}^b(\Sigma)$ is just the repetitive category of $\mathrm{coh}(\mathbb{X})$. Moreover, $\mathcal{H}$ has almost split sequences and the Serre functor $\tau : \mathcal{H} \longrightarrow \mathcal{H}$ serves as Auslander-Reiten translation. Denote by τ^- the inverse Auslander-Reiten translation.

0.3.10 (Grothendieck group). Denote by $\mathrm{K}_0(\mathbb{X})$ the Grothendieck group of $\mathcal{H}$. Since $\mathcal{H}$ and $\mathrm{mod}(\Sigma)$ have the same bounded derived category, we have $\mathrm{K}_0(\mathbb{X}) = \mathrm{K}_0(\Sigma)$, and this is a free abelian group of finite rank. We denote by $[X]$ the class in $\mathrm{K}_0(\mathbb{X})$ of an object $X \in \mathcal{H}$.

$\mathrm{K}_0(\mathbb{X})$ is equipped with the (normalized) Euler form $\langle -, - \rangle$. This bilinear form is defined on classes of objects X, Y in $\mathcal{H}$ by

$$\langle [X], [Y] \rangle = \frac{1}{m}\big(\dim_k \mathrm{Hom}_{\mathcal{H}}(X, Y) - \dim_k \mathrm{Ext}^1_{\mathcal{H}}(X, Y)\big),$$

where m is a positive integer such that the image of the resulting bilinear form generates $\mathbb{Z}$.

The Auslander-Reiten translation τ induces the *Coxeter transformation*, which we also denote by τ (by a slight abuse of notation), and which is an automorphism on $\mathrm{K}_0(\mathbb{X}) = \mathrm{K}_0(\Sigma)$ preserving the Euler form. The *radical* of $\mathrm{K}_0(\mathbb{X})$ is defined by $\mathrm{Rad}(\mathrm{K}_0(\mathbb{X})) = \{\mathbf{x} \in \mathrm{K}_0(\mathbb{X}) \mid \tau \mathbf{x} = \mathbf{x}\}$.

0.3.11 (Weights). For each $x \in \mathbb{X}$ let $p(x)$ be the rank of the tube $\mathcal{T}_x$. That is, $p(x)$ is the number of isomorphism classes of simple objects in $\mathcal{U}_x$. The tube $\mathcal{T}_x$, or the point x, is called *homogeneous* ([91]), if $p(x) = 1$, *exceptional* otherwise. $\mathbb{X}$ is called *homogeneous* if all $p(x) = 1$. Clearly, a point x is exceptional if and only if a simple object S_x in $\mathcal{U}_x$ is exceptional.

Each exceptional curve admits only a finite number of exceptional points. Denote by $x_1, \ldots, x_t \in \mathbb{X}$ the exceptional points. We call the numbers $p_i = p(x_i) > 1$ *weights*, accordingly $(p_1, \ldots, p_t)$ the *weight sequence*.

0.3.12 (Rank). We define the *rank* of sheaves: Let $x_0 \in \mathbb{X}$, and let S_0 be a simple sheaf in the tube $\mathcal{U}_{x_0}$ of rank p_0. Let $\mathbf{w} := \sum_{j=0}^{p_0-1}[\tau^j S_0]$, which is an element of $\mathrm{Rad}\,\mathrm{K}_0(\mathbb{X})$. By [70] we can assume that x_0 is a so-called rational point (see 0.4.4), that is, $\mathbb{Z}\mathbf{w}$ is a direct summand of $\mathrm{Rad}\,\mathrm{K}_0(\mathbb{X})$. After normalizing the linear form $\langle -, \mathbf{w} \rangle$ on $\mathrm{K}_0(\mathbb{X})$ by the factor $c := [\mathbb{Z} : \langle \mathrm{K}_0(\mathbb{X}), \mathbf{w} \rangle]$, we get a surjective linear form, compatible with the Coxeter transformation: For each $\mathbf{x} \in \mathrm{K}_0(\mathbb{X})$ define $\mathrm{rk}\,\mathbf{x} := \frac{1}{c}\langle \mathbf{x}, \mathbf{w} \rangle$, and moreover $\mathrm{rk}(X) = \mathrm{rk}([X])$ for each $X \in \mathcal{H}$. Let $X \in \mathcal{H}$ be indecomposable. Then $\mathrm{rk}(X) = 0$ if and only if $X \in \mathcal{H}_0$; if $X \in \mathcal{H}_+$, then $\mathrm{rk}(X) > 0$.

0.3.13 (Function field). The quotient category of $\mathcal{H}$ modulo the Serre subcategory $\mathcal{H}_0$, formed by the objects of finite length, is equivalent to the category of finite dimensional vector spaces over some skew field which is (up to isomorphism) uniquely determined by $\mathbb{X}$. We call this skew field the *function field*. We denote it by $k(\mathbb{X}) = k(\mathcal{H})$:

$$\mathcal{H}/\mathcal{H}_0 \simeq \mathrm{mod}(k(\mathbb{X})).$$

We call an exceptional curve $\mathbb{X}$ *commutative* if the function field $k(\mathbb{X})$ is commutative.

The function field is known to be of finite dimension over its centre and to be an algebraic function (skew) field of one variable over k (in the sense of [**106**]), see [**7**].

If $L \in \mathcal{H}_+$ is a *line bundle*, that is, of rank one, then $k(\mathbb{X})$ is isomorphic to the endomorphism ring of L considered as object in $\mathcal{H}/\mathcal{H}_0$ (given by fractions of morphisms of the same degree). Moreover, the rank of an object $X \in \mathcal{H}$ agrees with the dimension of the vector space over $k(\mathbb{X})$ corresponding to X considered as object in $\mathcal{H}/\mathcal{H}_0$.

The function field coincides with the endomorphism ring of the generic module associated with the separating tubular family $\mathrm{mod}_0(\Sigma)$ and was already studied in detail in [**90**].

0.3.14 (Special line bundle). From each of the exceptional tubes choose a simple sheaf $S_i \in \mathcal{U}_{x_i}$. Note that these simple sheaves are exceptional. In the following let $L \in \mathcal{H}_+$ be a line bundle, and assume additionally that for each $i \in \{1, \ldots, t\}$ we have $\mathrm{Hom}(L, \tau^j S_i) \neq 0$ if and only if $j \equiv 0 \bmod p_i$. Such a line bundle L exists by [**70**, Prop. 4.2] and is called *special*. It follows from [**70**, 5.2] that L is exceptional, since $\mathrm{End}_{\mathcal{H}}(L)$ is a skew field and $\mathbf{a} := [L]$ is a root in $\mathrm{K}_0(\mathbb{X})$. Recall from [**66**, **57**] that $\mathbf{v} \in \mathrm{K}_0(\mathbb{X})$ is a *root* if $\langle \mathbf{v}, \mathbf{v} \rangle > 0$ and $\frac{\langle \mathbf{v}, \mathbf{x} \rangle}{\langle \mathbf{v}, \mathbf{v} \rangle} \in \mathbb{Z}$ for all $\mathbf{x} \in \mathrm{K}_0(\mathbb{X})$. For example, the class of an exceptional object is a root. Moreover, an exceptional object is uniquely determined (up to isomorphism) by its class.

In the sequel, we will always consider $\mathcal{H}$ together with a special line bundle L, also called a *structure sheaf*. Of course, if $\mathbb{X}$ is homogeneous then each line bundle is special.

0.3.15 (Degree). Let p be the least common multiple of the weights $p_1, \ldots, p_t$. Define $\langle\!\langle -, - \rangle\!\rangle := \sum_{j=0}^{p-1} \langle \tau^j -, - \rangle$ and define the *degree* function $\deg : \mathrm{K}_0(\mathbb{X}) \longrightarrow \mathbb{Z}$ by

$$\deg \mathbf{x} := \frac{1}{c} \big(\langle\!\langle \mathbf{a}, \mathbf{x} \rangle\!\rangle - \mathrm{rk}\, \mathbf{x} \langle\!\langle \mathbf{a}, \mathbf{a} \rangle\!\rangle \big),$$

where as above $\mathbf{a} = [L]$.

0.3.16 (Underlying tame bimodule). Let L be a special line bundle and $S_1, \ldots, S_t$ simple objects from the different exceptional tubes such that $\mathrm{Hom}(L, S_i) \neq 0$. Let $\mathcal{S} = \{\tau^j S_i \mid 1 \le i \le t,\ j \not\equiv -1 \bmod p_i\}$. Then the right perpendicular category $\mathcal{S}^{\perp}$ is equivalent to $\mathrm{mod}(\Lambda)$, where Λ is a tame hereditary k-algebra of the form

$$\Lambda = \begin{pmatrix} G & 0 \\ M & F \end{pmatrix},$$

where $M = {}_F M_G$ is a *tame bimodule* (also called affine bimodule), that is:

- F and G are skew fields, finite dimensional over k;
- k lies in the centres of F and G and acts centrally on M.
- For the dimensions, $[M : F] \cdot [M : G] = 4$;

We say that M is a (tame) bimodule of (*dimension*) *type* $(2,2)$, $(1,4)$ or $(4,1)$ if this pair is $([M : F], [M : G])$. We call the number $\varepsilon \in \{1, 2\}$ the *numerical type* of M (or of $\mathbb{X}$), which is defined by

$$\varepsilon = \begin{cases} 1 & \text{if } M \text{ is of type } (2,2). \\ 2 & \text{if } M \text{ is of type } (1,4) \text{ or } (4,1). \end{cases}$$

The numerical type is an invariant of the curve $\mathbb{X}$.

With $\kappa := \langle [L], [L] \rangle$, for the normalization factor $c = [\mathbb{Z} : \langle K_0(\mathbb{X}), \mathbf{w} \rangle]$ as above we have $c = \kappa\varepsilon$.

0.3.17 (Automorphism groups). Let $\mathbb{X}$ be an exceptional curve with associated abelian hereditary category $\mathcal{H}$ and structure sheaf L. Denote by $\mathrm{Aut}(\mathcal{H})$ the automorphism class group of $\mathcal{H}$, that is, the group of isomorphism classes of autoequivalences of $\mathcal{H}$ (in the literature sometimes also called the Picard group [**8**], which has a different meaning in our presentation). We call this group the *automorphism group* of $\mathcal{H}$ and call the elements *automorphisms*. (If there is need to emphasize the base field k, we also write $\mathrm{Aut}_k(\mathbb{X})$ and use a similar notation in analogue situations.)

By a slight abuse of terminology, we will also call the autoequivalences themselves automorphisms, that is, the representatives of such classes; if F is an autoequivalence, then its class in the automorphism group is also denoted by F.

The subgroup of elements of $\mathrm{Aut}(\mathcal{H})$ fixing L (up to isomorphism) is denoted by $\mathrm{Aut}(\mathbb{X})$, the automorphism group of $\mathbb{X}$. (We will later see that this group does not dependent on L.)

Each element $\phi \in \mathrm{Aut}(\mathcal{H})$ induces a bijective map $\overline{\phi}$ on the points of $\mathbb{X}$ by $\phi(\mathcal{U}_x) = \mathcal{U}_{\overline{\phi}(x)}$ for all $x \in \mathbb{X}$. We call $\overline{\phi}$ the *shadow* of ϕ. If ϕ lies in the kernel of the homomorphism $\mathrm{Aut}(\mathcal{H}) \longrightarrow \mathrm{Bij}(\mathbb{X})$, $\phi \mapsto \overline{\phi}$, then we call ϕ *point fixing* (or *invisible* on $\mathbb{X}$). If $\overline{\phi}(x) = x$ we also say (by a slight abuse of terminology) that the point x is fixed by ϕ. Similarly, if $\overline{\phi}(x) = y$ we also write $\phi(x) = y$.

Denote by $\mathrm{Aut}_0(\mathcal{H})$ the (normal) subgroup of $\mathrm{Aut}(\mathcal{H})$ given by the point fixing automorphisms.

Non-trivial elements of $\mathrm{Aut}(\mathbb{X})$ which are point fixing are called *ghost automorphisms*, or just *ghosts*. The subgroup $\mathcal{G}$ of $\mathrm{Aut}(\mathbb{X})$ formed by the ghosts is called the *ghost group*. It is a normal subgroup of $\mathrm{Aut}(\mathcal{H})$. We have $\mathcal{G} = \mathrm{Aut}(\mathbb{X}) \cap \mathrm{Aut}_0(\mathcal{H})$. We call the factor group $\mathrm{Aut}(\mathbb{X})/\mathcal{G}$ the *geometric automorphism group* of $\mathbb{X}$, its elements *geometric automorphisms*. By a slight abuse of terminology, we also call the elements in $\mathrm{Aut}(\mathbb{X})$ which are not ghosts *geometric*.

Denote by $\mathrm{Aut}(\mathrm{D}^b(\mathbb{X}))$ the group of isomorphism classes of exact autoequivalences of the triangulated category $\mathrm{D}^b(\mathbb{X})$, called the *automorphism group* of $\mathrm{D}^b(\mathbb{X})$. (Compare also [**9**]. There is also the related notion of the derived Picard group [**82**].)

0.3.18 (Projective coordinate algebras). Let H be a finitely generated abelian group of rank one, which is equipped with a partial order $\leq$, compatible with the group structure. Let $R = \bigoplus_{h \in H} R_h$ be an H-graded k-algebra, such that each homogeneous component R_h is finite dimensional over k and such that $R_h = 0$ for $0 \nleq h$. Assume moreover that R is a finitely generated k-algebra and noetherian. Note that we do not require that R is commutative.

Denote by $\mathrm{mod}^H(R)$ the category of finitely generated right H-graded R-modules, and by $\mathrm{mod}_0^H(R)$ the full subcategory of graded modules of finite length (which is equivalent to finite k-dimension). This is a Serre subcategory of $\mathrm{mod}^H(R)$, that is, it is closed under subobjects, quotients and extensions. The quotient category $\mathrm{mod}^H(R)/\mathrm{mod}_0^H(R)$ is taken in the Serre-Grothendieck-Gabriel sense. We refer to [85].

Then the graded algebra R is called *a projective coordinate algebra* for $\mathbb{X}$ if there is an equivalence of categories

$$\mathcal{H} \simeq \frac{\mathrm{mod}^H(R)}{\mathrm{mod}_0^H(R)}.$$

Each exceptional curve admits a projective coordinate algebra, even a $\mathbb{Z}$-graded one (see 6.2.1). Thus, in the terminology of [2], $\mathcal{H}$ is a (noncommutative) noetherian projective scheme.

Note that a projective coordinate algebra is not uniquely determined by $\mathbb{X}$. One of the main aims of this article is to show that there is a projective coordinate algebra with "good" ringtheoretical properties.

0.4. Tubular shifts

One of the most important concepts we will use in these notes is that of shift automorphisms as developed in [70], which is a particular class of tubular mutations [71, 79, 80]. For the details we refer to [70]. Since we will also deal with the degree shift of graded objects, we will call a shift automorphism in the sense of [70] a *tubular shift* or just *shift associated to a point*.

0.4.1. Let $\mathbb{X}$ be an exceptional curve with associated hereditary category $\mathcal{H}$ and tubular family $\mathcal{H}_0 = \coprod_{x \in \mathbb{X}} \mathcal{U}_x$, with connected uniserial length categories $\mathcal{U}_x$ which are pairwise orthogonal. We fix a point $x \in \mathbb{X}$ of weight $p(x)$. Let S_x be a simple object in $\mathcal{U}_x$, denote by $\mathcal{S}_x$ additive closure of the Auslander-Reiten orbit of S_x, which consists of the semisimple objects from $\mathcal{U}_x$.

Let M be an object. By the semisimplicity of the category $\mathcal{S}_x$, for the object

$$M_x = \bigoplus_{j=1}^{p(x)} \mathrm{Ext}^1(\tau^j S_x, M) \otimes_{\mathrm{End}(S_x)} \tau^j S_x$$

there is a natural isomorphism of functors

(0.4.1) $\eta_M : \mathrm{Hom}(-, M_x)|_{\mathcal{S}_x} \xrightarrow{\sim} \mathrm{Ext}^1(-, M)|_{\mathcal{S}_x},$

which by the Yoneda lemma can be viewed as short exact sequence

$$\eta_M : 0 \longrightarrow M \xrightarrow{\alpha_M} M(x) \xrightarrow{\beta_M} M_x \longrightarrow 0$$

such that the Yoneda composition $\mathrm{Hom}(U, M_x) \longrightarrow \mathrm{Ext}^1(U, M)$, $f \mapsto \eta_M \cdot f$ is an isomorphism for each $U \in \mathcal{S}_x$. η_M is called the $\mathcal{S}_x$-*universal extension* of M. (If $p(x) = 1$ we also call it $\mathcal{S}_x$-universal.) By means of the identification $\mathrm{Hom}(-, M_x)|_{\mathcal{S}_x} = \mathrm{Ext}^1(-, M)|_{\mathcal{S}_x}$ the assignment $M \mapsto M_x$ extends to a functor $u \mapsto u_x$ for each $u : M \longrightarrow N$ such that $u \cdot \eta_M = \eta_N \cdot u_x$. Then M_x (u_x) is called the *fibre* of M (of u, resp.) in x.

Similarly, let

$$_xM = \bigoplus_{j=1}^{p(x)} \mathrm{Hom}(\tau^j S_x, M) \otimes_{\mathrm{End}(S_x)} \tau^j S_x.$$

Then there is a natural isomorphism $\gamma_M : \mathrm{Hom}(-, {}_xM)|_{\mathcal{S}_x} \longrightarrow \mathrm{Hom}(-, S)|_{\mathcal{S}_x}$, which corresponds to a morphism $\gamma_M : {}_xM \longrightarrow M$, called $\mathcal{S}_x$-*universal*.

0.4.2. Denote by $\mathcal{N}_x$ the full subcategory of $\mathcal{H}$ consisting of objects M such that $\mathrm{Hom}(\mathcal{U}_x, M) = 0$. There is an autoequivalences $\sigma_x : \mathcal{H} \longrightarrow \mathcal{H}$ associated to the point x or the tube with index x, therefore called *tubular shift associated to* x, with the following properties:

(1) For each $M \in \mathcal{N}_x$ the object $\sigma_x(M)$ agrees with $M(x)$. Moreover, if also $N \in \mathcal{N}_x$ and $u \in \mathrm{Hom}(M, N)$, then $\sigma_x(u)$ agrees with the unique morphism $u(x)$ making the following diagram commutative

$$\begin{array}{ccccccccc}
0 & \longrightarrow & M & \xrightarrow{\alpha_M} & M(x) & \xrightarrow{\beta_M} & M_x & \longrightarrow & 0 \\
& & \downarrow{\scriptstyle u} & & \downarrow{\scriptstyle u(x)} & & \downarrow{\scriptstyle u_x} & & \\
0 & \longrightarrow & N & \xrightarrow{\alpha_N} & N(x) & \xrightarrow{\beta_N} & N_x & \longrightarrow & 0.
\end{array}$$

More precisely, $u(x)$ is already uniquely determined by commutativity of the left hand square.

(2) Let $0 \longrightarrow M \xrightarrow{f} M' \xrightarrow{g} C \longrightarrow 0$ be a short exact sequence such that $M, M' \in \mathcal{N}_x$ and $C \simeq M_x$. Then there is a commutative exact diagram

$$\begin{array}{ccccccccc}
0 & \longrightarrow & M & \xrightarrow{\alpha_M} & M(x) & \xrightarrow{\beta_M} & M_x & \longrightarrow & 0 \\
& & \| & & \uparrow{\scriptstyle \simeq} & & \uparrow{\scriptstyle \simeq} & & \\
0 & \longrightarrow & M & \xrightarrow{f} & M' & \xrightarrow{g} & C & \longrightarrow & 0.
\end{array}$$

(In fact, the isomorphism $\mathrm{Hom}(C, M_x) \simeq \mathrm{Ext}^1(C, M)$ implies the pullback diagram above. In this diagram, the map $C \longrightarrow M_x$ is monic, since its kernel is a subobject of $M' \in \mathcal{N}_x$. Since C and M_x have the same length, the map is also epic.)

(3) $M \in \mathcal{H}_+$ implies $\sigma_x(M) \in \mathcal{H}_+$.

(4) If $y \neq x$ then there is a natural isomorphism $\sigma_x \circ \sigma_y \simeq \sigma_y \circ \sigma_x$. On $\mathcal{U}_y$ the tubular shift σ_x acts functorially as the identity.

(5) If $M \in \mathcal{U}_x$ then there is the exact sequence

$$0 \longrightarrow {}_xM \xrightarrow{\gamma_M} M \longrightarrow \sigma_x(M) \longrightarrow M_x \longrightarrow 0.$$

σ_x acts on objects in $\mathcal{U}_x$ like τ^-.

(Remark: Assume that $x \in \mathbb{X}$ is homogeneous. Then it is not true in general that the tubular shift σ_x or the Auslander-Reiten translation τ coincides with the identity functor on the homogeneous tube $\mathcal{U}_x$. This will be shown in 5.4.2 and 5.4.3.)

(6) There is a natural transformation $\varepsilon_x : 1_{\mathcal{H}} \longrightarrow \sigma_x$, coinciding on $\mathcal{N}_x$ with α. This natural transformation is also denoted by $1_{\mathcal{H}} \xrightarrow{x} \sigma_x$.

(7) On $\mathrm{K}_0(\mathbb{X})$, σ_x induces the automorphism

$$\mathbf{y} \mapsto \mathbf{y} - \sum_{j=1}^{p(x)} \frac{\langle \mathbf{y}, [\tau^j S_x] \rangle}{|\mathrm{End}(S_x)|} [\tau^j S_x],$$

where $|-|$ denotes the dimension over k.

0.4.3 (Multiplicity [**90, 70**]). As special case we have: Let L be a special line bundle with $\mathrm{Ext}^1(S_x, L) \neq 0$. Then the S_x-universal extension of L has the form

$$0 \longrightarrow L \longrightarrow L(x) \longrightarrow S_x^{e(x)} \longrightarrow 0,$$

with $e(x) = [\mathrm{Ext}^1(S_x, L) : \mathrm{End}(S_x)]$. The number $e(x)$ is called the *multiplicity* of x. It does not dependent on the choice of the special line bundle L. By Serre duality, $e(x)$ coincides with $[\mathrm{Hom}(L, \tau S_x) : \mathrm{End}(S_x)]$. A point x is called *multiplicity free* if $e(x) = 1$. The exceptional curve $\mathbb{X}$ is called *multiplicity free* ([**90**]) if $e(x) = 1$ holds for all $x \in \mathbb{X}$.

0.4.4 (Index). With the notations as in the preceding number, the dimension $f(x) = \frac{1}{\varepsilon} \cdot [\mathrm{Ext}^1(S_x, L) : \mathrm{End}(L)]$ is called the *index* of x. (Recall, that ε denotes the numerical type of $\mathbb{X}$.) A point x is called *rational* if $f(x) = 1$. Such a point always exists [**70**, Prop. 4.1]. We call a homogeneous point x *unirational* if $e(x) = 1 = f(x)$. Such a point does not always exist, compare 0.6.1.

The product $e(x) \cdot f(x)$ is denoted by $d(x)$ and called the *exponent* of x.

0.4.5 (Symbol). Let $x_1, \ldots, x_t \in \mathbb{X}$ be the exceptional points with weights $p_i = p(x_i)$, and let $f_i = f(x_i)$ the index and $d_i = d(x_i)$ the exponent of the point x_i $(i = 1, \ldots, t)$. Let ε be the numerical type of $\mathbb{X}$. Following [**66**] we call the matrix

$$\sigma[\mathbb{X}] = \sigma_\infty[\mathbb{X}] = \left(\begin{array}{ccc|c} p_1, \ldots, p_t & \\ d_1, \ldots, d_t & \varepsilon \\ f_1, \ldots, f_t & \end{array} \right)$$

the *symbol* of $\mathbb{X}$. (We make the convention, that rows of the form $1, 1, \ldots, 1$ and the entry $\varepsilon = 1$ are omitted in the notation of the symbol.)

For a point $x \in \mathbb{X}$ we call the numbers $p(x)$, $f(x)$ and $e(x)$ (or $d(x)$) together also the *symbol data* of x. For any simple object S_x concentrated in x such that $\mathrm{Hom}(L, S_x) \neq 0$ we have $[\mathrm{Hom}(L, S_x) : k] = \varepsilon \cdot f(x) \cdot [\mathrm{End}(L) : k]$ and $[\mathrm{End}(S_x) : k] = \frac{\varepsilon \cdot f(x) \cdot [\mathrm{End}(L):k]}{e(x)}$. Moreover, $\deg(S_x) = f(x) \cdot \frac{p}{p(x)}$ with the least common multiple p of $p_1, \ldots, p_t$.

The symbol of $\mathbb{X}$ determines the Grothendieck group $\mathrm{K}_0(\mathbb{X})$ uniquely up to isomorphism which preserves the Euler form. (The converse also holds if $\mathbb{X}$ is domestic.) We refer to [**66, 57**].

0.4.6 ([**70**, S15]). Let $M, N \in \mathcal{H}_+$ be non-zero and $x \in \mathbb{X}$ a point. Then for sufficiently large (positive) n,

  (a) $\mathrm{Hom}(M, \sigma_x^n(N)) \neq 0$.
  (b) $\mathrm{Hom}(\sigma_x^n(M), N) = 0$.

By Serre duality, one gets similar formulae for the extension spaces.

0.4.7 (The Picard group). Denote by $\mathrm{Pic}(\mathbb{X})$ the subgroup of $\mathrm{Aut}(\mathcal{H})$ generated by all tubular shifts σ_x $(x \in \mathbb{X})$ and call it the *Picard group*. It is always abelian. By $\mathrm{Pic}_0(\mathbb{X})$ denote the subgroup of those elements of $\mathrm{Pic}(\mathbb{X})$ of degree zero. That is, $\sigma \in \mathrm{Pic}(\mathbb{X})$ is of degree zero if and only if the degree of $\sigma(L)$ is zero. By 0.4.5, $\deg(L(x)) = d(x) \cdot p/p(x)$ for all $x \in \mathbb{X}$, and it follows, that the definition does not depend on the choice of the structure sheaf L. Every torsion element from $\mathrm{Pic}(\mathbb{X})$ lies in $\mathrm{Pic}_0(\mathbb{X})$. (The converse is an open question in general.)

0.4.8. Let $x \in \mathbb{X}$ be a point and $\phi \in \text{Aut}(\mathcal{H})$. The object $\phi(S_x)$ is simple, concentrated in a point $y \in \mathbb{X}$. Then as elements in $\text{Aut}(\mathcal{H})$,

$$\sigma_y = \phi \circ \sigma_x \circ \phi^{-1}.$$

In particular, $\text{Pic}(\mathbb{X})$ is a normal subgroup of $\text{Aut}(\mathcal{H})$.

It is sometimes useful to have a stronger formulation: there is a natural isomorphism $\sigma_y \circ \phi \xrightarrow{\mu} \phi \circ \sigma_x$, which is compatible with the natural transformations $\phi \xrightarrow{\phi x_?} \phi \sigma_x$ and $\phi \xrightarrow{y_{\phi(?)}} \sigma_y \phi$. One shows this by first considering for $M \in \mathcal{N}_x$ the ϕ-image of the $\mathcal{S}_x$-universal extension of M on the one hand and the $\mathcal{S}_y$-universal extension of $\phi(M)$ on the other hand, and then using 0.4.2 (2). In a second step the natural isomorphism on $\mathcal{N}_x$ obtained in this way will be extended to $\mathcal{H}$. (Compare the proof of 3.1.2 for further details.)

0.5. Tame bimodules and homogeneous exceptional curves

0.5.1 (Bimodule algebra). In these notes we only consider bimodules $M = {}_F M_G$, where F and G are skew fields of finite dimension over k, with k lying in their centres, and such that M is finite dimensional over k, with k acting centrally. Such bimodules, finite dimensional over a central subfield, are also called algebraic [89].

Each bimodule $M = {}_F M_G$ gives rise to a finite dimensional k-algebra

$$\Lambda = \begin{pmatrix} G & 0 \\ M & F \end{pmatrix}$$

which is hereditary. Moreover, this algebra is of tame representation type (that is, not of finite and not of wild type) if and only if M is a tame bimodule, that is, $[M : F] \cdot [M : G] = 4$. In this case, the indecomposable regular modules lie in homogeneous tubes.

More precisely, in the tame case there is the trisection

$$\text{mod}(\Lambda) = \mathcal{P} \vee \mathcal{R} \vee \mathcal{Q},$$

that is, $\text{mod}(\Lambda)$ is the additive closure of $\mathcal{P}$, $\mathcal{R}$ and $\mathcal{Q}$, where $\mathcal{P}$ is the preprojective component, consisting of the Auslander-Reiten orbits of two indecomposable projective modules, dually $\mathcal{Q}$ is the preinjective component and

$$\mathcal{R} = \coprod_{x \in \mathbb{X}} \mathcal{U}_x$$

consists of the regular modules, whose indecomposable summands lie in homogeneous tubes $\mathcal{U}_x$, and thus $\mathcal{R} = \text{mod}_0(\Lambda)$ is the separating tubular family.

0.5.2 (Associated hereditary category). In this tame bimodule case the associated hereditary abelian k-category $\mathcal{H}$ is constructed in a simple manner,

$$\mathcal{H} = \mathcal{Q}[-1] \vee \mathcal{P} \vee \mathcal{R},$$

formed inside the bounded derived category $\text{D}^b(\text{mod}(\Lambda))$. In other words, $\mathcal{H}$ is obtained from $\text{mod}(\Lambda)$ by shifting the preinjective component to the left and gluing it to the preprojective component, creating thus a category without non-zero projectives or injectives.

$\mathcal{H}$ is a hereditary category on which the Auslander-Reiten translation gives rise to an autoequivalence and which admits a tilting object with endomorphism ring Λ. Hence $\mathcal{H}$ is nothing else but a homogeneous exceptional curve. Conversely, any homogeneous exceptional curve is obtained in this way.

The regular Λ-modules become the objects of finite length in $\mathcal{H}$. The objects of $\mathcal{Q}[-1] \vee \mathcal{P}$ are the vector bundles. Denote by L a fixed line bundle (which means that the corresponding preprojective (or preinjective) Λ-module is of rank 1 (-1, respectively). The line bundle L plays the role of the structure sheaf.

0.5.3 (Homogeneous exceptional curves). A homogeneous exceptional curve $\mathcal{H}$ is characterized by the following properties:

- $\mathcal{H}$ is a connected small abelian k-category with finite dimensional morphism and extension spaces.
- $\mathcal{H}$ is hereditary and noetherian and contains no non-zero projective object.
- $\mathcal{H}$ admits a tilting object.
- For each simple object $S \in \mathcal{H}$ we have $\mathrm{Ext}^1(S, S) \neq 0$.

The last condition precisely means that all tubes in $\mathcal{H}_0$ are homogeneous.

0.5.4 (Structure sheaf, tilting bundle). In the homogeneous case any line bundle is special and can play the role of the structure sheaf L. Let L be fixed. Let $\overline{L} \in \mathcal{H}_+$ be indecomposable such that there is an irreducible morphism $L \longrightarrow \overline{L}$. Then $M = \mathrm{Hom}(L, \overline{L})$ is the (up to duality unique) underlying tame bimodule of $\mathbb{X}$, and $\mathrm{rk}(\overline{L}) = \varepsilon$ is the numerical type. Moreover, $T = L \oplus \overline{L}$ is a tilting bundle such that $\Lambda = \mathrm{End}(T)$ is as in 0.5.1.

0.5.5 (The centre). Let M be a tame bimodule and Λ and $\mathcal{H}$ as above. Since the centre of $\mathcal{H}$ (or of Λ) is a field, it is sometimes useful to assume – without loss of generality – that k is the centre of Λ. But we will not assume this in general.

The centre of M is defined to be the set of all pairs $(f, g) \in F \times G$ such that $fm = mg$ for all $m \in M$. In this case, f belongs to the centre of F and g to the centre of G, and the centre of M is a field K which can be identified with its projections into F or into G, see [89, 5.2]. Of course, K/k is a finite field extension, M is a tame bimodule over K, and Λ is a tame hereditary K-algebra with centre K.

Concerning dual bimodules there is the following general fact.

0.5.6 (Dual bimodule [22, 2.1.1]). Let $_F M_G$ be an F-G-bimodule over k. There are isomorphisms of G-F-bimodules

$$\mathrm{Hom}_F(_F M_G, {}_F F_F) \simeq \mathrm{Hom}_k(_F M_G, k) \simeq \mathrm{Hom}_G(_F M_G, {}_G G_G).$$

0.5.7. The index set $\mathbb{X}$ above is naturally equipped with geometric structure, given by the hereditary category $\mathcal{H}$. It is the aim of the first part to study this structure. Whereas for algebraically closed field $\overline{k}$ this structure is well understood ($\mathbb{X}$ is the projective line $\mathbb{P}^1(\overline{k})$) the structure in general can be very complicated. Unlike in the algebraically closed case it is in general even impossible to determine all the points of $\mathbb{X}$ explicitly. Also, there are points of many different kinds (of different degrees, having non-isomorphic endomorphism skew fields of the associated skyscraper sheaves,...)

One of the still easiest examples is the Kronecker algebra over $k = \mathbb{Q}$. Already in this innocent looking example the explicit structure of $\mathbb{X}$ is quite complicated. The points of $\mathbb{X} = \mathbb{P}^1(\mathbb{Q})$ (in the scheme-sense) are in one-to-one correspondence with the irreducible homogeneous polynomials in $\mathbb{Q}[X, Y]$ (up to multiplication with non-zero scalars). In this ring there are infinitely many irreducible homogeneous polynomials in any degree. It is hopeless to classify all these irreducible elements. *Any* finite field extension of $\mathbb{Q}$ occurs as endomorphism ring of a skyscraper sheaf.

0.6. Rational points

Let $\mathbb{X}$ be an exceptional curve. In general it is hopeless to know all points of $\mathbb{X}$. But often one has some control over the points "lying on the lowest level", the rational points. Recall that for an exceptional curve $\mathbb{X}$ rational points x, that means, with $f(x) = 1$, always exist (see 0.4.4).

Let $M = {}_F M_G$ be a $(2,2)$-bimodule with associated homogeneous exceptional curve $\mathbb{X}$ and m a non-zero element in M. Then the representation

$$S_x = (F_F, \, G_G, \, \pi_m : F_F \otimes {}_F M_G \simeq M_G \longrightarrow M_G/mG \simeq G_G),$$

involving the canonical projection, induces a simple object in $\mathcal{H}$ concentrated in some point $x \in \mathbb{X}$ (see [29]). Obviously, x is a rational point, and each rational point arises in this way.

The following lemma is taken from [53, C.1]. It is a very useful tool for calculating multiplicities of rational points.

LEMMA 0.6.1. *Let $M = {}_F M_G$ be a $(2,2)$-bimodule. Let m be a non-zero element in M. Let $x \in \mathbb{X}$ be the induced rational point. For the multiplicity we have*

$$e(x) = \frac{[F : k]}{[(Fm \cap mG) : k]}.$$

If M is a simple bimodule, or more generally, if m is a bimodule generator of M, then $e(x) > 1$.

PROOF. We have $\mathrm{Ker}(\pi_m) = 1 \otimes mG$. Any endomorphism of S_x is given by the commutative diagram

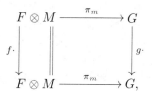

with $f \in F$ and $g \in G$, and it follows, that $fmG \subset mG$. Consider the subring $R = \{f \in F \mid fmG \subset mG\}$ of F. Sending $(f\cdot, g\cdot)$ to f yields an isomorphism $\mathrm{End}(S_x) \simeq R$: Injectivity follows by applying π_m to an element $1 \otimes y$, with $y \in M \setminus mG$. Surjectivity follows, since the map $f\cdot$ for $f \in R$ restricts to the kernel of π_m, and hence induces a morphism $g\cdot$ on G. Moreover, $f \mapsto fm$ gives rise to an isomorphism $R \simeq Fm \cap mG$, and with $e(x) = \frac{[\mathrm{Hom}(L, S_x):k]}{[\mathrm{End}(S_x):k]} = \frac{[\mathrm{End}(L):k]}{[\mathrm{End}(S_x):k]}$ the stated formula follows. Finally, if m is a bimodule generator, we have $R \neq F$, since $M \neq mG$.

Alternative proof. We can consider m as monomorphism between L and $\overline{L}$ (irreducible map) with cokernel $S = S_x$. Lifting endomorphisms from $\mathrm{End}(S)$ to $\mathrm{End}(\overline{L})$ induces an isomorphism between $\mathrm{End}(S)$ and the subskewfield of $\mathrm{End}(\overline{L})$ of those elements $g \in \mathrm{End}(\overline{L})$ such that there is an $f \in \mathrm{End}(L)$ such that $g \circ m = m \circ f$. $\square$

COROLLARY 0.6.2. *Let $M = {}_F M_G$ be a $(2,2)$-bimodule. Then M is a non-simple bimodule if and only if there exists a unirational point x.*

PROOF. If M is simple then no unirational point exists by 0.6.1. If M is non-simple then $F \simeq G$, without loss of generality $F = G$, and then ${}_F M = F \oplus F$.

By [**89**] there is an automorphism of F over k and an $(\alpha, 1)$-derivation δ of F such that $(x, y) \cdot f = (xf + y\delta(f), y\alpha(f))$ for all f, x, $y \in F$. Let $m = (1, 0)$. Then $Fm = mF$, hence by 0.6.1 the induced rational point has multiplicity one. $\qquad \square$

LEMMA 0.6.3. *Let $M = {}_F M_G$ be a $(2, 2)$-bimodule. Let m and m' be non-zero elements in M inducing points x and x', respectively. Then $x = x'$ if and only if there are non-zero elements $f \in F$ and $g \in G$ such that $m' = fmg$.*

PROOF. Assume $x = x'$. Let $S = S_x$. Consider the exact sequences

$$0 \longrightarrow L \xrightarrow{\ m\ } \overline{L} \xrightarrow{\ p\ } S \longrightarrow 0$$

$$0 \longrightarrow L \xrightarrow{\ m'\ } \overline{L} \xrightarrow{\ p'\ } S \longrightarrow 0,$$

Applying the functor $\mathrm{Hom}(\overline{L}, -)$ to the lower sequence, since $\mathrm{Ext}^1(\overline{L}, L) = 0$ there is some $f \in \mathrm{End}(\overline{L})$ such that $p' \circ f = p$, and this proves one direction. The converse is trivial. $\qquad \square$

Part 1

The homogeneous case

CHAPTER 1

Graded factoriality

In this chapter we show how to associate with each homogeneous exceptional curve $\mathbb{X}$ a (not necessarily commutative) graded factorial domain; it will be shown in the next chapter that such a factorial domain is a projective coordinate algebra for $\mathbb{X}$. We use the term "graded factorial" for a graded version of rings which are called noncommutative noetherian unique factorization rings by Chatters and Jordan [**13, 47**].

Such a coordinate algebra will be constructed as orbit algebra $\Pi(L, \sigma)$ where L is a line bundle and σ a so-called efficient automorphism on $\mathcal{H}$. This means that σ is point fixing such that the cyclic group $\langle \sigma \rangle$ acts on the set of line bundles "as transitively as possible". This condition guarantees that the middle term in each S_x-universal extension (defined in 0.4.1) of L

$$0 \longrightarrow L \overset{\pi_x}{\longrightarrow} L(x) \longrightarrow S_x^{e(x)} \longrightarrow 0$$

satisfies $L(x) \simeq \sigma^d(L)$ for some natural number d depending on x, and therefore the kernel π_x can be interpreted as a homogeneous element in the orbit algebra $\Pi(L, \sigma)$. Note that this orbit algebra is noncommutative in general.

It is not difficult to see that for each homogeneous exceptional curve an efficient automorphism exists. The main result of this chapter is the following theorem (see 1.2.3 and 1.5.1).

THEOREM. *Let* $R = \Pi(L, \sigma)$ *with* σ *being efficient. Let* S_x *be a simple sheaf concentrated in the point* $x \in \mathbb{X}$. *Let*

$$0 \longrightarrow L \overset{\pi_x}{\longrightarrow} \sigma^d(L) \longrightarrow S_x^e \longrightarrow 0$$

be the S_x-*universal extension of* L. *Then the following conditions hold*
  (1) *The element* π_x *is normal, that is,* $R\pi_x = \pi_x R$.
  (2) $P_x = R\pi_x$ *is a homogeneous prime ideal.*
  (3) P_x *is a completely* homogeneous *prime ideal (that is,* R/P_x *is a graded domain) if and only if* $e = 1$.
  Moreover, for any homogeneous prime ideal P *of height one there is a point* $x \in \mathbb{X}$ *such that* $P = P_x$.

Because of the last statement and since R is also a noetherian domain, we say that R is graded factorial, in analogy to commutative algebra.

With the theorem we have established a link between tubular shifts and the (projective) prime spectrum of R. It turns out that graded factoriality is very useful for studying the geometry of $\mathbb{X}$.

1.1. Efficient automorphisms

Let $\mathbb{X}$ be a homogeneous exceptional curve with associated hereditary category $\mathcal{H}$.

1.1.1. Recall that $\mathrm{Aut}_0(\mathcal{H})$ is the subgroup of $\mathrm{Aut}(\mathcal{H})$ consisting of those automorphisms (autoequivalences) ϕ which are point fixing, that is, which satisfy $\phi(S_x) \simeq S_x$ for all $x \in \mathbb{X}$. Note that for example the Auslander-Reiten translation τ, its inverse τ^- and all tubular shifts are in $\mathrm{Aut}_0(\mathcal{H})$. We will usually assume (without loss of generality) that a point fixing automorphism σ of $\mathbb{X}$ (that is, a ghost) satisfies $\sigma(A) = A$ (equality) for all objects $A \in \mathcal{H}_0$.

1.1.2. We fix a line bundle L ("structure sheaf"). Then L determines the degree function such that $\deg(L) = 0$ (see 0.3.15). There is an indecomposable $\overline{L} \in \mathcal{H}_+$ such that there is an irreducible map $L \longrightarrow \overline{L}$. Then $T = L \oplus \overline{L}$ is a tilting bundle on $\mathcal{H}$ such that $\Lambda = \mathrm{End}(T)$ is a tame hereditary bimodule algebra over k and the $\mathrm{End}(\overline{L})$-$\mathrm{End}(L)$-bimodule $M = \mathrm{Hom}(L, \overline{L})$ serves as underlying tame bimodule. The rank of $\overline{L}$ coincides with the numerical type ε of M, hence is one or two. The Auslander-Reiten quiver (species) of $\mathcal{H}_+$ has the following shape:

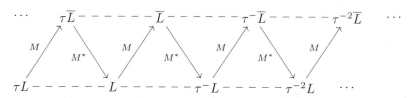

where the dotted lines indicate the Auslander-Reiten orbits and M^* denotes the dual bimodule of M. A line bundle is (up to isomorphism) uniquely determined by its degree. The precise value depends on whether $\overline{L}$ is a line bundle or not. If it is a line bundle (that is, $\varepsilon = 1$) then we have $\deg(\tau^{-n}L) = 2n$ and $\deg(\tau^{-n}\overline{L}) = 2n+1$ for all $n \in \mathbb{Z}$. If it is not a line bundle (that is, $\varepsilon = 2$), then $\deg(\tau^{-n}L) = n$ for all $n \in \mathbb{Z}$.

DEFINITION 1.1.3 (Efficient automorphism). Let $\sigma \in \mathrm{Aut}(\mathcal{H})$. We call σ *efficient* if it is point fixing and such that $\deg(\sigma L) > 0$ is minimal with this property.

Obviously, if σ is efficient and γ is a ghost automorphism, then $\gamma \circ \sigma$ and $\sigma \circ \gamma$ are efficient.

LEMMA 1.1.4. *Let $\mathbb{X}$ be a homogeneous exceptional curve. Then there exists an efficient automorphism σ. Moreover, such an automorphism σ is uniquely determined up to a ghost automorphism.*

PROOF. For the existence it is sufficient to remark that the inverse Auslander-Reiten translation τ^- is point fixing with $\deg(\tau^-L) = 2/\varepsilon > 0$. Thus there is an efficient automorphism σ such that $1 \leq \deg(\sigma L) \leq 2/\varepsilon$. (Moreover, either $\sigma(L) \simeq \overline{L}$ or $\sigma(L) \simeq \tau^-(L)$.) If σ' is also efficient then $\sigma^{-1} \circ \sigma'$ fixes all objects in $\mathcal{H}$ and hence is a ghost automorphism. $\square$

1.1.5 (The orbit cases). Here we present a division of tame bimodules. Each tame bimodule M belongs to precisely one of the following three classes, called *orbit cases*:

    I M is a tame bimodule of type $(1,4)$ or $(4,1)$. In this case, the set of all line bundles coincides with the Auslander-Reiten orbit of L and also with the $\mathrm{Aut}_0(\mathcal{H})$-orbit of L.

    II M is a tame bimodule of type $(2,2)$ and there is precisely one $\mathrm{Aut}_0(\mathcal{H})$-orbit of line bundles, that is, $\mathrm{Aut}_0(\mathcal{H})$ acts transitively on the set of all line bundles.

    III M is a tame bimodule of type $(2,2)$, and there are precisely two $\mathrm{Aut}_0(\mathcal{H})$-orbits of line bundles, coinciding with the Auslander-Reiten orbits.

Denote by $\mathcal{O}$ the $\mathrm{Aut}(\mathcal{H})$-orbit and by $\mathcal{O}_0$ the $\mathrm{Aut}_0(\mathcal{H})$-orbit of L, that is, $F \in \mathcal{H}$ lies in $\mathcal{O}_0$ if and only if there is $\sigma \in \mathrm{Aut}_0(\mathcal{H})$ such that $\sigma(L) \simeq F$. (Similarly for $\mathcal{O}$.)

REMARK 1.1.6. In orbit cases I and III the inverse Auslander-Reiten translation $\sigma = \tau^-$ is an efficient automorphism. In orbit case II there is by definition a $\sigma \in \mathrm{Aut}_0(\mathcal{H})$ such that $\sigma(L) \simeq \overline{L}$, which gives an efficient automorphism. Moreover, by comparing dimensions of homomorphism spaces, $\sigma(\overline{L}) \simeq \tau^- L$ follows. Thus, in all orbit cases, if σ is efficient, the cyclic group $\langle \sigma \rangle$ acts transitively on $\mathcal{O}_0$.

DEFINITION 1.1.7. Let $\sigma \in \mathrm{Aut}(\mathcal{H})$. We call σ

- *positive*, if $\deg(\sigma L) > 0$.
- *exhaustive*, if the cyclic group $\langle \sigma \rangle$ acts transitively on $\mathcal{O}_0$.
- *transitive*, if $\langle \sigma \rangle$ acts transitively on $\mathcal{O}$.

LEMMA 1.1.8. *An autoequivalence $\sigma \in \mathrm{Aut}(\mathcal{H})$ is efficient if and only if it is positive, point fixing and exhaustive.*

PROOF. Follows immediately from Remark 1.1.6 by considering each of the three orbit cases. $\qquad\square$

The following consequence is the main reason for defining efficient automorphisms and will be used in the next section.

COROLLARY 1.1.9. *Let σ be efficient and σ_x be a tubular shift associated to a point x. Then there is some positive integer d such that $\sigma_x(L) \simeq \sigma^d(L)$.* $\qquad\square$

REMARK 1.1.10. (1) Assume that the underlying tame bimodule is non-simple of type $(2,2)$. Then there is a unirational point $x_0 \in \mathbb{X}$. Let σ_0 be the corresponding tubular shift. Then $\langle \sigma_0 \rangle$ acts transitively on the set of isomorphism classes of line bundles, implying orbit case II.

(2) If k is algebraically closed, or if $k = \mathbb{R}$, or if k is a finite field, then each tame bimodule is either of orbit case I or non-simple as in (1).

(3) The bimodule $M = {}_{\mathbb{Q}(\sqrt{2})}\mathbb{Q}(\sqrt{2}, \sqrt{3})_{\mathbb{Q}(\sqrt{3})}$ belongs to orbit case III. More generally each $(2,2)$-bimodule ${}_F M_G$ with non-isomorphic F and G belongs to this class since there is no automorphism sending L to $\overline{L}$.

(4) If k is algebraically closed then an efficient automorphism σ is uniquely determined. If $k = \mathbb{R}$ the same is true unless $M = \mathbb{C} \oplus \overline{\mathbb{C}}$; in that case we have the two possibilities $\sigma = \sigma_0$ as in (1) and $\sigma = \sigma_0 \circ \gamma = \gamma \circ \sigma_0$, where γ is induced by complex conjugation. This will be proved in Section 5.3.

(5) In cases I and II a tubular shift σ_x at a point x is exhaustive (hence efficient) if and only if x is a unirational point. In case III a tubular shift σ_x is exhaustive if and only if either $f(x) = 1$ and $e(x) = 2$ or $f(x) = 2$ and $e(x) = 1$. Tubular shifts which are efficient do not always exist, see Example 1.1.13 below.

(6) There are also simple bimodules in orbit case II, see Example 1.1.13 below.

(7) We will see that in orbit case III for any point the product $e(x) \cdot f(x)$ is even (see 1.2.1 or 1.6.6).

The division of tame bimodules into the orbit cases will be very useful in the following. On the other hand, this division is quite formal. In order to get a better understanding it would be interesting to solve the following problem.

PROBLEM 1.1.11. Find a criterion from which one can easily decide whether a given $(2, 2)$-bimodule is of orbit case II or III.

We will later see that efficient automorphisms which are tubular shifts are advantageous for our considerations (see for example 1.7.1). In general efficient tubular shifts do not exist, see Example 1.1.13 below.

PROBLEM 1.1.12. Does there always exist an efficient automorphism lying in the Picard group?

EXAMPLE 1.1.13. In [**29**, 5.3] the following type of $(2, 2)$-bimodules is considered. Let K be a (commutative) field with subfield F and G, each of index 2 such that $k = F \cap G$ is of finite index m in F (and G). Then let M be the F-G-bimodule K.

(1) M is a simple bimodule if and only if $F \neq G$. (Let $F \neq G$. Assume that N is a proper, non-trivial subbimodule of M. Then for $0 \neq n \in N$ we have $Fn = N = nG$. Take an element $f \in F \setminus G$ to get a contradiction.)

(2) If m is odd then M is of orbit case II. This follows from 0.6.1 together with 1.1.10 (7).

(3) If m is odd and $F \neq G$ (hence $m > 1$) then there is no efficient tubular shift. This follows from 1.1.10 (5) with 0.6.1.

(4) An explicit example for which (3) holds is given as follows: Let $k = \mathbb{Q}$ and $M = {}_{\mathbb{Q}(\sqrt[3]{2})}\mathbb{Q}(\sqrt[3]{2}, \zeta)_{\mathbb{Q}(\zeta\sqrt[3]{2})}$ (where ζ is a primitive third root of unity). By 0.6.1 each rational point x has multiplicity $e(x) = 3$.

1.1.14. If an efficient automorphism σ is fixed, for an object $A \in \mathcal{H}$ and $n \in \mathbb{Z}$ we write $A(n) = \sigma^n(A)$. A similar notation for morphisms is used. Note in this context that one can actually assume that σ is not only an autoequivalence but an automorphism, that is, an invertible functor (see the discussion in [**2**]).

We fix a line bundle L. Then each line bundle in $\mathcal{O}_0$ is (up to isomorphism) of the form $L(n)$ for some (unique) $n \in \mathbb{Z}$. Let ℓ be the degree of $L(1)$; then $\ell = 1$ in orbit cases I and II, and $\ell = 2$ in orbit case III. We call ℓ the *orbit number* of $\mathbb{X}$ or of M.

1.1.15 (Orbit algebra). Let σ be an efficient automorphism. Let R be the orbit algebra

$$\Pi(L, \sigma) = \bigoplus_{n \geq 0} \mathrm{Hom}_{\mathcal{H}}(L, L(n)),$$

where the multiplication is defined for elements $r \in \mathrm{Hom}(L, L(m))$ and $s \in \mathrm{Hom}(L, L(n))$ by $s * r = \sigma^m(s) \circ r$. (We use the symbol $*$ only if we would like to emphasize that this orbit algebra multiplication is meant.) This yields a positively $\mathbb{Z}$-graded k-algebra with $R_0 = \mathrm{End}(L)$ a skew field, and all homogeneous components R_n are finite dimensional over k. Since non-zero morphisms between line bundles are monomorphisms, R is a *graded domain*, that is, for all non-zero homogeneous a, $b \in R$ we have $ab \neq 0$.

For each $n \geq 0$ we have

$$\dim_{R_0} R_n = \begin{cases} n+1 & \text{orbit case II} \\ 2n+1 & \text{orbit cases I and III.} \end{cases}$$

In one formula: $\dim_{R_0} R_n = \varepsilon \ell n + 1$. Since each morphism between vector bundles is a sum of compositions of irreducible morphisms, R is generated in degrees 0 and 1. (This is also true if σ is not exhaustive.)

Note that for example in case $M = {}_{\mathbb{Q}(\sqrt{2})}\mathbb{Q}(\sqrt{2}, \sqrt{3})_{\mathbb{Q}(\sqrt{3})}$ the orbit algebras $\Pi(L, \sigma)$ and $\Pi(\overline{L}, \sigma)$ are not isomorphic as graded algebras.

For a $\mathbb{Z}$-graded algebra, we call the functor $\mathrm{Mod}^{\mathbb{Z}}(R) \longrightarrow \mathrm{Mod}^{\mathbb{Z}}(R)$, $X \mapsto X(1)$, the degree shift. For $R = \Pi(L, \sigma)$ this is induced by the automorphism σ, which we therefore also call the *degree shift* (see 2.1.6).

1.2. Prime ideals and universal extensions

We keep the notations from the previous section. In particular, σ is efficient.

1.2.1. Let S be a simple object concentrated in the point $x \in \mathbb{X}$. Let $e = e(x)$ be the multiplicity and $f = \deg(S)$ the degree of S (compare 0.4.5). With the orbit number ℓ from 1.1.14 the quotient $d := ef/\ell$ is a natural number. It is easy to see (compare 1.1.9) that then $L(x) \simeq L(d)$ (that is, $\sigma_x(L) \simeq \sigma^d(L)$), and therefore the S-universal extension of L is given by

$$0 \longrightarrow L \xrightarrow{\pi} L(d) \longrightarrow S^e \longrightarrow 0$$

and the kernel π is a homogeneous element in $R = \Pi(L, \sigma)$ of degree d.

LEMMA 1.2.2. *Let σ be a positive and point fixing automorphism. Denote $\sigma^n(F)$ by $F(n)$. Consider the following diagram of exact sequences*

$$0 \longrightarrow F_2 \xrightarrow{f_2} F_1 \xrightarrow{f_1} F_0 \longrightarrow 0$$
$$\downarrow u_0$$
$$0 \longrightarrow G_2 \xrightarrow{g_2} G_1 \xrightarrow{g_1} G_0 \longrightarrow 0$$

with F_1, F_2, G_1, $G_2 \in \mathcal{H}_+$ and F_0, $G_0 \in \mathcal{H}_0$. Then there is some integer $n \geq 0$ and a commutative diagram

$$0 \longrightarrow F_2 \xrightarrow{f_2} F_1 \xrightarrow{f_1} F_0 \longrightarrow 0$$
$$u_2 \downarrow \qquad u_1 \downarrow \qquad u_0 \downarrow$$
$$0 \longrightarrow G_2(n) \xrightarrow{g_2(n)} G_1(n) \xrightarrow{g_1(n)} G_0 \longrightarrow 0$$

PROOF. For $n \geq 0$ apply $\mathrm{Hom}(F_1, -)$ to the short exact sequence

$$0 \longrightarrow G_2(n) \xrightarrow{g_2(n)} G_1(n) \xrightarrow{g_1(n)} G_0 \longrightarrow 0.$$

Since by 0.4.6 for sufficiently large n we have

$$\mathrm{Ext}^1(F_1, G_2(n)) \simeq \mathrm{D}\,\mathrm{Hom}(G_2(n), \tau F_1) = 0,$$

the map $\mathrm{Hom}(F_1, G_1(n)) \longrightarrow \mathrm{Hom}(F_1, G_0)$ is surjective, and the assertion follows immediately.   □

Of course, the lemma can be generalized in an obvious way to the weighted case.

Let P be a (two-sided) homogeneous ideal in R. Then P is called a (homogeneous) *prime ideal*, if for all a, b homogeneous, $aRb \subset P$ implies $a \in P$ or $b \in P$. Moreover, P is called a (homogeneous) *completely prime ideal*, if for all a, $b \in R$ homogeneous, $ab \in P$ implies $a \in P$ or $b \in P$. A homogeneous element a in R is called *normal* if $Ra = aR$. We additionally assume that normal and central elements are non-zero. If R is a graded domain then a normal element a defines a graded algebra automorphism γ_a on R by $ra = a\gamma_a(r)$ for all $r \in R$, and a is central if and only if $\gamma_a = 1$.

THEOREM 1.2.3. *Let $R = \Pi(L, \sigma)$ with σ being efficient. Let S_x be a simple sheaf concentrated in the point $x \in \mathbb{X}$. Let $e = e(x)$ be the multiplicity, $f = \deg(S_x)$ the degree of S_x, $d = ef/\ell$ and*

$$0 \longrightarrow L \xrightarrow{\pi_x} L(d) \longrightarrow S_x^e \longrightarrow 0$$

the S_x-universal extension of L. Then the following holds.
  (1) *The homogeneous element π_x is normal.*
  (2) *$P_x = R\pi_x$ is a homogeneous prime ideal.*
  (3) *P_x is a completely homogeneous prime ideal if and only if $e = 1$.*

PROOF. We drop the index x and write $S = S_x$, $\pi = \pi_x$ and $P = P_x$.

(1) Let $r \in R$ be homogeneous of degree n. We have the commutative diagram with (universal) exact sequences

$$
\begin{array}{ccccccccc}
0 & \longrightarrow & L & \xrightarrow{\ \pi\ } & L(d) & \longrightarrow & S^e & \longrightarrow & 0 \\
  & & {\scriptstyle r}\downarrow & & {\scriptstyle s}\downarrow & & {\scriptstyle r_x}\downarrow & & \\
0 & \longrightarrow & L(n) & \xrightarrow{\ \pi(n)\ } & L(n+d) & \longrightarrow & S^e & \longrightarrow & 0,
\end{array}
$$

for some s (by universality 0.4.2 (1)). Since σ is an equivalence there is some homogeneous $t \in R$ such that $s = t(d)$. Then, by the definition of the multiplication in R we get $\pi r = t\pi$. Hence $\pi R \subset R\pi$. The reverse inclusion follows since each homogeneous component is finite dimensional.

(2) Let P' be the graded ideal in R, whose homogeneous elements are given by those r such that $r_x = 0$, where r_x is given by the following diagram

$$
\begin{array}{ccccccccc}
0 & \longrightarrow & L & \xrightarrow{\ \pi\ } & L(d) & \longrightarrow & S^e & \longrightarrow & 0 \\
  & & {\scriptstyle r}\downarrow & & {\scriptstyle r'}\downarrow & & {\scriptstyle r_x}\downarrow & & \\
0 & \longrightarrow & L(n) & \xrightarrow{\ \pi(n)\ } & L(d+n) & \longrightarrow & S^e & \longrightarrow & 0,
\end{array}
$$

(r homogeneous of degree n) where r' is given as in (1) by universality. Obviously, $\pi \in P'$. Moreover, if $r \in P'$ is homogeneous of degree n, then there is a commutative, exact diagram

$$
\begin{array}{ccccccccccc}
0 & \longrightarrow & L & \xrightarrow{\ \pi\ } & L(d) & \xrightarrow{\ p\ } & S^e & \longrightarrow & 0 \\
  & & {\scriptstyle r}\downarrow & & {\scriptstyle r'}\downarrow & & {\scriptstyle r_x=0}\downarrow & & \\
0 & \longrightarrow & L(n) & \xrightarrow{\ \pi(n)\ } & L(d+n) & \xrightarrow{\ p(n)\ } & S^e & \longrightarrow & 0.
\end{array}
$$

Hence the zero morphism $0 : S^e \longrightarrow L(d+n)$ satisfies $r_x = p(n) \circ 0$. By the Homotopy-Lemma [**45**, Lemma B.1] (applied to this special situation) there is an $s \in \mathrm{Hom}(L(d), L(n))$ such that $r = s\pi$. Hence $P' = R\pi$ follows.

Note that $\mathrm{End}(S^e) \simeq \mathrm{M}_e(D)$ (where $D = \mathrm{End}(S)$ is a skew field) is a prime ring and σ induces an automorphism of this ring. Using the formula $(s*r)_x = \sigma^m(s_x) \circ r_x$ (where m is the degree of r) it is sufficient to show that for some $n \geq 0$ the map

$$R_n \longrightarrow \mathrm{End}(S^e), \ r \mapsto r_x$$

is surjective. But this follows from Lemma 1.2.2.

(3) Let $e = 1$. If a, $b \in R$ are homogeneous such that $ab \in P$ then $(ab)_x = 0$. Since $\mathrm{End}(S)$ is a skew field, $a_x = 0$ or $b_x = 0$, hence $a \in P$ or $b \in P$ and P is completely prime. For the converse, if $e > 1$, then there are non-zero matrices A, $B \in \mathrm{M}_e(\mathrm{End}\,S)$ such that $A \cdot B = 0$. By the proof of (2) there are homogeneous a, $b \in R$ such that $b_x = B$ and $a_x = \sigma^{-m}(A)$ (where m is the degree of b). It follows that $ab \in P$, but $a \notin P$ and $b \notin P$. Hence P is not completely prime. □

1.3. Prime ideals as annihilators

In this section we give another description of the homogeneous prime ideals which occur in Theorem 1.2.3. We assume $R = \Pi(L, \sigma)$, where σ is efficient. As usual, we set $F(n) = \sigma^n(F)$ for all $F \in \mathcal{H}$.

1.3.1 (Fibre map). Let S be simple, concentrated in x, let $e = e(x)$. For an $f \in \mathrm{Hom}(L, L')$, where L' is some line bundle, we have the following commutative diagram with universal exact sequences

$$
\begin{array}{ccccccccc}
0 & \longrightarrow & L & \overset{\pi}{\longrightarrow} & L(d) & \longrightarrow & S^e & \longrightarrow & 0 \\
& & \downarrow{f} & & \downarrow{f'} & & \downarrow{f_x} & & \\
0 & \longrightarrow & L' & \overset{\pi'}{\longrightarrow} & L'(d) & \longrightarrow & S^e & \longrightarrow & 0,
\end{array}
$$

with fibre map f_x.

1.3.2 (1-irreducible maps). Let f be a (non-zero) morphism between line bundles. Then f is called 1-*irreducible*, if whenever $f = gh$ with morphisms g and h between line bundles, then g or h is an isomorphism. The following facts are obvious:

(1) Each non-zero map between line bundles has a factorization into 1-irreducible maps.

(2) A morphism between line bundles is 1-irreducible if and only if its cokernel is a simple object.

(3) Each simple object is cokernel of a 1-irreducible map. Moreover, one of the line bundles can be chosen arbitrarily.

(4) If $u : L \longrightarrow L(n)$ is a 1-irreducible map, then it is an irreducible element in R. The converse does not hold in general, in orbit case III.

The following lemma is a fundamental statement on 1-irreducible maps.

LEMMA 1.3.3. *Let S be simple, concentrated in x, let $\pi = \pi_x$ and $e = e(x)$, and let*

$$0 \longrightarrow L \overset{u}{\longrightarrow} L' \longrightarrow S \longrightarrow 0$$

be exact, where L' is a line bundle. Then there is a morphism $v \in \operatorname{Hom}(L', L(d))$ such that $\pi = vu$. Moreover, the fibre $u_x : S^e \longrightarrow S^e$ of u has kernel and cokernel isomorphic to S.

PROOF. Since $\operatorname{Hom}(S, S^e) \simeq \operatorname{Ext}^1(S, L)$ by (0.4.1), there is a commutative exact diagram

$$
\begin{array}{ccccccccc}
0 & \longrightarrow & L & \overset{\pi}{\longrightarrow} & L(d) & \longrightarrow & S^e & \longrightarrow & 0 \\
& & \| & & \uparrow & & \uparrow & & \\
0 & \longrightarrow & L & \overset{u}{\longrightarrow} & L' & \longrightarrow & S & \longrightarrow & 0,
\end{array}
$$

which proves the first part. For the fibre maps we have $0 = \pi_x = v_x u_x$. If u_x would be an epimorphism, then we would have $v_x = 0$. Then, as in the proof of part (2) of Theorem 1.2.3 we would get a non-zero $s \in \operatorname{Hom}(L'(d), L(d)) = 0$, contradiction. Hence, the cokernel of u_x is non-zero, hence isomorphic to S. By the snake lemma, the same follows for the kernel. $\qquad \square$

PROPOSITION 1.3.4. *Let S be simple, concentrated in the point x, and P be the corresponding homogeneous prime ideal (by Theorem 1.2.3). Let M be the graded left R-module $\bigoplus_{n \geq 0} \operatorname{Ext}^1_{\mathcal{H}}(S, L(n))$. Then $P = \operatorname{Ann}_R(M)$.*

PROOF. Let $r \in R_n$, $r \neq 0$. The S-universal extension induces a commutative exact diagram

$$
\begin{array}{ccccccccc}
0 & \longrightarrow & L & \overset{\pi}{\longrightarrow} & L(d) & \longrightarrow & S^e & \longrightarrow & 0 \\
& & {\scriptstyle r}\downarrow & & \downarrow & & \| & & \\
0 & \longrightarrow & L(n) & \longrightarrow & X & \longrightarrow & S^e & \longrightarrow & 0.
\end{array}
$$

If $r \in \operatorname{Ann}_R(M)$ then the lower sequence splits and $r \in P$ follows immediately. For the converse, we show more generally the next proposition. $\qquad \square$

PROPOSITION 1.3.5. *Let S be a simple sheaf concentrated in x and $P = R\pi$ be the corresponding homogeneous prime ideal. For each $n \in \mathbb{N}$ let $S^{(n)}$ be the indecomposable sheaf of length n with socle S. Let $M^{(n)}$ be the graded left R-module $\bigoplus_{i \geq 0} \operatorname{Ext}^1(S^{(n)}, L(i))$. Then $\operatorname{Ann}_R(M^{(n)}) \supset R\pi^n = P^n$.*

PROOF. There is a short exact sequence

$$
0 \longrightarrow S \overset{i}{\longrightarrow} S^{(n)} \overset{p}{\longrightarrow} S^{(n-1)} \longrightarrow 0,
$$

which induces a short exact sequence of graded modules

$$
0 \longrightarrow M^{(n-1)} \overset{p^*}{\longrightarrow} M^{(n)} \overset{i^*}{\longrightarrow} M \longrightarrow 0.
$$

We have to show that $\pi M^{(n)} \subset p^*(M^{(n-1)}) \ (\simeq M^{(n-1)})$. A homogeneous, non-zero element η in $M^{(n)}$ induces the following pushout diagram

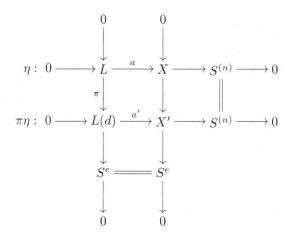

If X decomposes, $X = L' \oplus E$ with $E \neq 0$ of finite length and L' a line bundle, then $E \simeq S^{(i)}$ for some $1 \leq i \leq n$ (since $S^{(n)}$ is uniserial), and we get the following commutative exact diagram

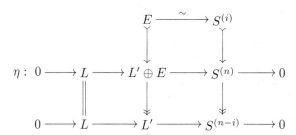

Then $\eta \in p^*(M^{(n-1)})$ follows. Similarly, if X' decomposes, then $\pi\eta \in p^*(M^{(n-1)})$ follows. But if X and X' are indecomposable, hence line bundles, then the middle vertical short exact sequence is (up to shift) the S-universal sequence (for L or for $\overline{L}$). It then follows that a_x is an isomorphism. Since a is a product of n morphisms between line bundles with cokernel S, we get a contradiction by Lemma 1.3.3. $\square$

COROLLARY 1.3.6. *For each $x \in \mathbb{X}$ denote by P_x the homogeneous prime ideal as in Theorem 1.2.3. For each infinite subset U of $\mathbb{X}$,*

$$\bigcap_{x \in U} P_x = 0.$$

PROOF. Denote by S_x the simple sheaf concentrated in x. Let $r \in R$ be non-zero and homogeneous of degree n. Choose $x \in U$ such that the cokernel in the short exact sequence

$$0 \longrightarrow L \xrightarrow{r} L(n) \longrightarrow C \longrightarrow 0,$$

has no non-zero summand which is concentrated in x. Denote by M_x the graded R-module $\oplus_{n \geq 0} \mathrm{Ext}^1(S_x, L(n))$. Then $r \notin \mathrm{Ann}_R(M_x) = P_x$ follows by the Homotopy-Lemma. $\square$

1.4. Noetherianness

Let $R = \Pi(L, \sigma)$, where σ is positive. We show that R is noetherian. Since this basic property is very important we give a detailed proof. The arguments are mainly taken from [**7**].

1.4.1. Let $\mathcal{L}$ be the full subcategory of $\mathcal{H}$ formed by all $L(n)$ (where $n \in \mathbb{Z}$). Denote by $\mathcal{L}_+$ the full subcategory formed by all $L(n)$ (where $n \geq 0$). Similarly, denote by $\mathcal{L}_-$ the full subcategory formed by all $L(n)$ (where $n \leq 0$). We denote by $\mathrm{Mod}\,\mathcal{L}$ the category of (covariant) k-functors $F : \mathcal{L} \longrightarrow \mathrm{Mod}(k)$ (similarly for $\mathcal{L}_+$ and $\mathcal{L}_-$), and by $\mathrm{Mod}^{\mathbb{Z}}(R)$ the category of $\mathbb{Z}$-graded right R-modules, by $\mathrm{Mod}^{\mathbb{Z}+}(R)$ those graded modules M with $M_n = 0$ for $n < 0$. The following obvious lemmas are proved as in [**7**, 3.6].

Lemma 1.4.2 (covariant functors = left graded modules). *There is an equivalence of k-categories*

$$\mathrm{Mod}(\mathcal{L}_+) \longrightarrow \mathrm{Mod}^{\mathbb{Z}+}(R^{op}), \quad F \mapsto \bigoplus_{n \geq 0} F(L(n)).$$

For $X \in \mathcal{H}$ denote by $(X, -]$ the functor $\mathrm{Hom}_{\mathcal{H}}(X, -)|_{\mathcal{L}_+}$, by $\mathrm{Ext}^1_{\mathcal{H}}(X, -]$ the functor $\mathrm{Ext}^1_{\mathcal{H}}(X, -)|_{\mathcal{L}_+}$.

Lemma 1.4.3 (contravariant functors = right graded modules). *There is an equivalence of k-categories*

$$\mathrm{Mod}(\mathcal{L}_-^{op}) \longrightarrow \mathrm{Mod}^{\mathbb{Z}+}(R), \quad F \mapsto \bigoplus_{n \geq 0} F(L(-n)).$$

For $X \in \mathcal{H}$ denote by $[-, X)$ the (contravariant) functor $\mathrm{Hom}_{\mathcal{H}}(-, X)|_{\mathcal{L}_-}$, by $\mathrm{Ext}^1_{\mathcal{H}}[-, X)$ the functor $\mathrm{Ext}^1_{\mathcal{H}}(-, X)|_{\mathcal{L}_-}$.

Proposition 1.4.4. $R = \Pi(L, \sigma)$ *is (graded) noetherian (left and right).*

Proof. (1) (compare [**7**, 4.2]) We have to show that the functor $(L, -]$ is noetherian. For this it is sufficient to show, that $(L, -]/U$ is noetherian for each non-zero cyclic subfunctor U. Let $n \in \mathbb{Z}$ such that there is an epimorphism

$$(L(n), -] \longrightarrow U.$$

Then there is a non-zero morphism

$$\eta : (L(n), -] \longrightarrow (L, -]$$

induced by some morphism $f : L \longrightarrow L(n)$. We get a short exact sequence

$$0 \longrightarrow L \overset{f}{\longrightarrow} L(n) \longrightarrow C \longrightarrow 0,$$

where $C \in \mathcal{H}_0$, which induces an exact sequence

$$0 \longrightarrow (C, -] \longrightarrow (L(n), -] \longrightarrow (L, -] \longrightarrow \mathrm{Ext}^1_{\mathcal{H}}(C, -].$$

Hence it suffices to show that the functor $\mathrm{Ext}^1_{\mathcal{H}}(S, -]$ is noetherian for each simple object $S \in \mathcal{H}_0$. But this follows precisely as in [**7**, 4.1]: we have to show that $\mathrm{Ext}^1(S, -]/U$ is noetherian for each non-zero cyclic (or finitely generated) subfunctor U. Let $(L(m), -] \longrightarrow \mathrm{Ext}^1(S, -]$ be a non-zero morphism with image U. This is induced by some non-split short exact sequence

$$0 \longrightarrow L(m) \longrightarrow L' \longrightarrow S \longrightarrow 0,$$

with some line bundle L'. We get an exact sequence

$$0 \longrightarrow (L', -] \longrightarrow (L(m), -] \longrightarrow \mathrm{Ext}^1(S, -] \longrightarrow \mathrm{Ext}^1(L', -].$$

By Serre duality and by the positivity of the grading of R we see, that the functor $\mathrm{Ext}^1(L', -]$ is of finite length, therefore the same holds true for $\mathrm{Ext}^1(S, -]/U$.

(2) Similarly, we have to show that the contravariant functor $[-, L)$ is noetherian. As above, it suffices to show that $[-, L)/V$ is noetherian for each non-zero cyclic (contravariant) subfunctor V. Let $n \in \mathbb{Z}$ be such that there is an epimorphism

$$[-, L(-n)) \longrightarrow V.$$

Then there is a non-zero morphism

$$[-, L(-n)) \longrightarrow [-, L)$$

induced by some morphism $g : L(-n) \longrightarrow L$. We get a short exact sequence

$$0 \longrightarrow L(-n) \xrightarrow{g} L \longrightarrow B \longrightarrow 0,$$

with $B \in \mathcal{H}_0$, hence an exact sequence

$$0 \longrightarrow [-, L(-n)) \longrightarrow [-, L) \longrightarrow [-, B).$$

It suffices to show that $[-, S)$ is noetherian for each simple sheaf. Again we show that $[-, S)/V$ is noetherian for each non-zero cyclic (contravariant) subfunctor V. We get a morphism

$$[-, L(m)) \longrightarrow [-, S)$$

with image V. This induces a short exact sequence

$$0 \longrightarrow L' \longrightarrow L(m) \longrightarrow S \longrightarrow 0,$$

with some line bundle L', hence an exact sequence

$$0 \longrightarrow [-, L') \longrightarrow [-, L(m)) \longrightarrow [-, S) \longrightarrow \mathrm{Ext}^1[-, L').$$

By Serre duality $\mathrm{Ext}^1[-, L')$ is of finite length, hence we get the result. $\square$

PROPOSITION 1.4.5. $R = \Pi(L, \sigma)$ has graded (classical) Krull dimension two. That is, since R is a graded local domain, the only homogeneous prime ideals are the zero ideal, the homogeneous maximal (left) ideal $\mathfrak{m} = \bigoplus_{n \geq 1} R_n$ and the homogeneous prime ideals of height one.

PROOF. Let $\mathbb{F} = \mathrm{mod}(\mathcal{L}_+)$, and denote by $\mathbb{F}_0$ the full subcategory of $\mathbb{F}$ of objects of finite length. Denote by $\mathbb{F}_1$ the Serre subcategory of objects in $\mathbb{F}$ which become of finite length in $\mathbb{F}/\mathbb{F}_0$. One shows as in [**7**, 4.6] that for each $n \in \mathbb{Z}$ the functor $(L(n), -]$ becomes simple in $\mathbb{F}/\mathbb{F}_1$. It follows (compare [**77**, 6.4.5]), that the classical Krull dimension of R is two. $\square$

1.5. Prime ideals of height one are principal

Let $\mathbb{X}$ be a homogeneous exceptional curve. For $x \in \mathbb{X}$ denote by P_x the homogeneous prime ideal of height 1 as in Theorem 1.2.3.

THEOREM 1.5.1. Let $R = \Pi(L, \sigma)$ with σ being efficient. Let P be a homogeneous prime ideal in R of height one. Then $P = P_x$ for some $x \in \mathbb{X}$.

PROOF. Let $a \in P$ be a non-zero homogeneous element. Let U be the subfunctor of $(L, -]$ corresponding to the graded module Ra. There is an epimorphism $(L(m), -] \longrightarrow U$. This induces a short exact sequence

$$(1.5.1) \qquad\qquad 0 \longrightarrow L \overset{a}{\longrightarrow} L(m) \longrightarrow C \longrightarrow 0,$$

where C is a coproduct $\coprod_{i=1}^{t} S_i^{(n_i)}$, with (not necessarily non-isomorphic) simple $S_1, \ldots, S_t$ in $\mathcal{H}_0$, concentrated in $x_1, \ldots, x_t$, respectively. Let $M_i^{(n_i)}$ be the graded left R-module $\bigoplus_{n \geq 0} \operatorname{Ext}^1(S_i^{(n_i)}, L(n))$ and $P_i = R\pi_i$ be the homogeneous prime ideal corresponding to x_i. We have $P_i^{n_i} \subset \operatorname{Ann}_R(M_i^{(n_i)})$. By applying $(-, -]$ to the short exact sequence (1.5.1) we get an exact sequence

$$0 \longrightarrow (L(m), -] \longrightarrow (L, -] \longrightarrow \operatorname{Ext}^1(C, -].$$

Thus R/Ra embeds into $\bigoplus_{i=1}^{t} M_i^{(n_i)}$. Hence we get (using 1.3.5)

$$P = \operatorname{Ann}_R(R/P) \supset \operatorname{Ann}_R(R/Ra) \supset \operatorname{Ann}_R\left(\bigoplus_{i=1}^{t} M_i^{(n_i)}\right) \supset \bigcap_{i=1}^{t} R\pi_i^{n_i}.$$

It follows that the product of some powers of the normal elements $\pi_1, \ldots, \pi_t$ is in P, hence $\pi_i \in P$ for some i. But then $P = P_i$ follows. $\qquad\square$

Combining Theorem 1.2.3 and Theorem 1.5.1 we get the following important result.

COROLLARY 1.5.2. *There is a natural bijection $x \mapsto R\pi_x$ between points $x \in \mathbb{X}$ and homogeneous prime ideals $P \subset R$ of height one, given by forming universal extensions. Under this bijection points of multiplicity one correspond to homogeneous completely prime ideals of height one.* $\qquad\square$

Invoking Krull's principal ideal theorem, a *commutative* noetherian integral domain is factorial (that is, a unique factorization domain) if and only if each prime ideal of height one is principal ([**76**, 20.1]). Inspired by this we make the following definition (where we allow abelian grading groups):

DEFINITION 1.5.3. Let R be a noetherian graded domain. Then R is called a (noncommutative) graded *factorial domain* if each homogeneous prime ideal of height one is principal, generated by a normal element.

In our setting a graded factorial domain is nothing but the graded version of a noncommutative noetherian unique factorisation ring (UFR) in the sense of Chatters and Jordan [**13**, **47**], where we here restrict our considerations to domains.

Note that on the other hand (besides the grading) the concept of a noncommutative unique factorization domain (UFD) in [**12**], where only completely prime ideals of height one are considered, is too restrictive for our purposes. (Actually we will show that the rings occurring in our setting are (graded) UFD's in the sense of Chatters [**12**] only when they are commutative, see Theorem 1.2.3 (3) and Theorem 4.3.5.) In his definition of unique factorization domains Cohn [**14**, **16**] focussed on irreducible elements (atoms) rather than on prime elements/ideals.

Corollary 1.5.2 has the following important consequence.

COROLLARY 1.5.4. *The orbit algebra $\Pi(L, \sigma)$, where σ is efficient, is a (noncommutative) graded factorial domain.* $\qquad\square$

COROLLARY 1.5.5. *In orbit cases I and III, the small preprojective algebra* $\Pi(L, \tau^-)$ *is graded factorial.*

PROOF. In these cases the inverse Auslander-Reiten translation τ^- is efficient. □

PROBLEM 1.5.6. It is an interesting question whether it is true that the completion $\widehat{R}$ of the graded factorial $R = \Pi(L, \sigma)$ is factorial again. This would provide a class of examples of (noncommutative) complete factorial rings. So far only very few examples are known (see [**107**]).

1.6. Unique factorization

Let $\mathbb{X}$ be homogeneous. Let $R = \Pi(L, \sigma)$ be an orbit algebra where σ is an efficient automorphism.

1.6.1 (Prime and irreducible elements). Recall that a non-zero, homogeneous element $a \in R$ is called *normal* if $Ra = aR$ holds. We call a (non-zero) homogeneous element π in R *prime* if it is normal and if it generates a homogeneous prime ideal (necessarily of height one). A non-zero homogeneous element u is called *irreducible* if $u = ab$ with a, b homogeneous, implies that either a or b is a unit.

We have a weak form of *Euclid's Lemma*: *If π is prime and divides the product ab, where a and b are homogeneous, such that a or b is normal, then π divides a or b.*

LEMMA 1.6.2. *Let π_1, π_2 be two prime elements. Then there is an $\alpha \in R_0^*$ such that $\pi_1 \pi_2 = \alpha \pi_2 \pi_1$.*

PROOF. This follows from the properties of universal extensions, but there is also a purely ring theoretical proof. The assertion is clear if $R\pi_1 = R\pi_2$. Assume that $R\pi_1 \neq R\pi_2$. Since π_1, π_2 are normal, we have $\pi_1 \pi_2 = \gamma(\pi_2)\pi_1$, where γ is a degree preserving automorphism on R. Then $\gamma(\pi_2)$ is also prime and π_2 divides $\gamma(\pi_2)\pi_1$, hence π_2 divides $\gamma(\pi_2)$, thus there is an $\alpha \in R_0^*$ such that $\gamma(\pi_2) = \alpha \pi_2$. □

PROPOSITION 1.6.3. *Let $R = \Pi(L, \sigma)$ with σ being efficient. Each non-zero normal element of R is a product of prime elements. This factorization is unique up to permutation and multiplication with units.*

PROOF. Let $a \neq 0$ be a normal non-unit. By the principal ideal theorem (see [**77**, 4.1.11]) there is a prime ideal P_1 of height one such that $a \in P_1$. There is a prime element π_1 such that $P_1 = R\pi_1$. Hence there is a homogeneous element r_1 such that $a = r_1 \pi_1$. Since r_1 is normal (see below), the assertion follows because of the positivity of the grading by induction. The uniqueness (up to a unit) follows from the preceding lemma.

We show that r_1 is normal. For each normal element f let $\gamma_f : R \longrightarrow R$ be the automorphism such that $sf = f\gamma_f(s)$ for each $s \in R$. Let $s \in R$, $r := r_1$, $\pi := \pi_1$. Then
$$(sr)\pi = s(r\pi) = sa = a\gamma_a(s) = r\pi\gamma_a(s) = r\gamma_\pi^{-1}\gamma_a(s)\pi,$$
hence $sr = r\gamma_\pi^{-1}\gamma_a(s)$, hence $Rr \subset rR$. The converse inclusion follows by $rs = \gamma_a^{-1}\gamma_\pi(s)r$ for each $s \in R$. □

COROLLARY 1.6.4 (Almost commutativity of normal elements). *Let f_1, f_2 be two normal elements. Then there is a homogeneous unit $\alpha \in R_0^*$ such that $f_1 f_2 = \alpha f_2 f_1$.* □

One can summarize the preceding results by saying that the non-zero normal elements modulo units form a UF-monoid in the sense of [**16**, Ch. 3], see also [**47**]. Moreover, as in [**47**, Prop. 2.2] it follows that each non-zero homogeneous ideal contains a normal element. (R is said to be (graded) conformal.)

Obviously, each homogeneous element $r \in R$, $r \neq 0$, is a product of irreducible elements. If a normal element is irreducible it is prime.

1.6.5 (Ringtheoretic meaning of $e(x)$ and $f(x)$). Let u be a homogeneous element of R with cokernel S. Obviously, if S is simple, then u is irreducible. The converse also holds in orbit cases I and II, that is, here irreducible elements and 1-irreducible maps are the same concept. Hence, with the notations of Theorem 1.2.3, in orbit cases I and II the prime element $\pi = \pi_x$ is a product of $e = e(x)$ irreducible elements: $\pi = u_1 u_2 \ldots u_e$. Moreover, each u_i is of degree $f = f(x) = \deg(S)$. $\square$

In orbit case III we get a slightly different result, since then there are irreducible elements (even of degree one in R) with cokernel of length two; they may have two different points as support.

PROPOSITION 1.6.6. *With the same notations as in Theorem 1.2.3, assume orbit case III. Then the following holds for $\pi = \pi_x$.*

(1) *If $f = \deg(S)$ is even, then $\pi = u_1 u_2 \ldots u_e$, where all $u_i \in R$ are irreducible of degree $f/2$ with cokernels isomorphic to S.*

(2) *If $f = \deg(S)$ is odd, then e is even and $\pi = u_1 u_2 \ldots u_t$, where $t = e/2$ and all $u_i \in R$ are irreducible of degree f with cokernels isomorphic to S^2.*

PROOF. A chain of projections $S^e \twoheadrightarrow S^{e-1} \twoheadrightarrow \ldots \twoheadrightarrow S$ yields a factorization $\pi = v_1 v_2 \ldots v_e$ with $v_i : L^{(e-i)} \longrightarrow L^{(e-i+1)}$, where $L^{(i)}$ are line bundles such that $L^{(0)} = L$ and $L^{(e)} = L(d)$. Moreover, the cokernels of the v_i are isomorphic to S, hence $\deg L^{(i+1)} = \deg L^{(i)} + f$. If f is even, all $L^{(i)}$ lie in the same orbit, whereas in case f is odd, precisely the $L^{(i)}$, where i is even, lie in the same orbit as L. So, in the first case, all $u_i := v_i$ are in R and irreducible. In the second case, the elements $u_i = v_{2i-1} v_{2i}$ are in R and irreducible. $\square$

Note that in orbit case III there are also irreducible elements in R with indecomposable cokernel $S_x^{(2)}$ of length two.

The following is a modification of Lemma 1.3.3 so that there is some left and right symmetry:

LEMMA 1.6.7. *Let S be simple, concentrated in x, and let $u \in R$ be irreducible such that*

$$0 \longrightarrow L \xrightarrow{u} L(f) \longrightarrow S \longrightarrow 0$$

is exact. Then there is a morphism $v' \in R_{d-f}$ such that $\pi_x = uv'$.

PROOF. Let $\pi = \pi_x = ab$ be a product of homogeneous elements. Then the cokernel of a (or b, resp.) is of the form S^f for some $0 \leq f \leq d$. Then, by universality, there is a b' such that $\pi = b'a$, and since $b'\pi = \pi b$, we get $b' = \gamma(b)$, where $\gamma : R \longrightarrow R$ is the automorphism such that $\pi r = \gamma(r)\pi$ for each $r \in R$. Now apply Lemma 1.3.3 to $\gamma^{-1}(u)$. $\square$

REMARK 1.6.8. There is also a version for irreducibles with cokernel S^2.

The same argument shows:

LEMMA 1.6.9. *Let $\pi = u_1 \dots u_t$ be a factorization of the prime element π into irreducibles $u_1, \dots, u_t$. Let $\gamma : R \longrightarrow R$ be the automorphism such that $\pi r = \gamma(r)\pi$ for all $r \in R$. Then*

$$
\begin{aligned}
\pi &= \gamma(u_t)u_1 \dots u_{t-1} &&= u_t\gamma^{-1}(u_1)\dots\gamma^{-1}(u_{t-1}) \\
&= \gamma(u_{t-1})\gamma(u_t)u_1 \dots u_{t-2} &&= u_{t-1}u_t\gamma^{-1}(u_1)\dots\gamma^{-1}(u_{t-2}) \\
&\ \ \vdots \qquad\qquad \vdots \\
&= \gamma(u_2)\dots\gamma(u_t)u_1 &&= u_2\dots u_t\gamma^{-1}(u_1).
\end{aligned}
$$

Accordingly, for all i and j there is an irreducible decomposition of π where u_i appears at position j. $\qquad\square$

We remark that the behaviour of irreducible elements is not well-understood. For example, it is not true in general, that a prime element remains the same under each permutation of its irreducible factors. It may happen, that under some permutation of the factors the cokernel is not semisimple. Moreover, in orbit case III it is even unknown whether each irreducible element is divisor of a normal element.

EXAMPLE 1.6.10. Let $\beta = e^{2\pi i/3}$ the third root of unity, let $k = \mathbb{Q}(\beta)$ and $K = k(t)$ with $t = \sqrt[3]{2}$. Let $R = K[X; Y, \alpha]$, where α is the k-automorphism on K given by $t \mapsto \beta t$. (We will see in the next section that R can be realized as orbit algebra $\Pi(L, \sigma)$ over a suitable exceptional curve with efficient automorphism σ.) Then

$$\pi = Y^3 - 2X^3 = (Y - tX)(Y - \beta tX)(Y - \beta^2 tX)$$

is a factorization of the central prime element π into irreducibles. Then

$$(Y - \beta tX)(Y - tX)(Y - \beta^2 tX) = Y^3 - 3\beta tXY^2 + 3\beta t^2 X^2 Y - 2X^3,$$

and the cokernel of this element is not semisimple, since it is not associated to π.

1.7. Examples of graded factorial domains

In this section we discuss some classes of examples of orbit algebras $\Pi(L, \sigma)$ with efficient σ. It follows in particular that these explicitly given algebras are graded factorial domains.

One main reason for preferring efficient automorphism which are tubular shifts (if they exist) when forming the orbit algebra is the following simple fact (which is some converse of Theorem 3.1.2 below). It ensures the existence of central prime elements of degree one which is very useful for calculating examples (see also the proof of Theorem 4.3.5).

LEMMA 1.7.1. *Let $R = \Pi(L, \sigma_x)$ be the orbit algebra defined by an efficient tubular shift σ_x at x. Then the prime element π_x associated with x is central in R.*

PROOF. Write $\sigma = \sigma_x$ and $\pi = \pi_x$. Since for all homogeneous elements $r \in R$ of degree $n \geq 0$ we have the commutative universal diagram

$$
\begin{array}{ccccccccc}
0 & \longrightarrow & L & \overset{\pi}{\longrightarrow} & \sigma L & \longrightarrow & S_x^{e(x)} & \longrightarrow & 0 \\
& & \downarrow{\scriptstyle r} & & \downarrow{\scriptstyle \sigma r} & & \downarrow{\scriptstyle r_x} & & \\
0 & \longrightarrow & \sigma^n L & \overset{\sigma^n \pi}{\longrightarrow} & \sigma^{n+1} L & \longrightarrow & S_x^{e(x)} & \longrightarrow & 0,
\end{array}
$$

the element π is central. $\qquad\square$

The non-simple bimodule case.

1.7.2. Let $\mathbb{X}$ be a homogeneous exceptional curve and $M = {}_F M_G$ be the underlying tame bimodule which we assume to be of type $(2,2)$ and to be non-simple. We identify $F = G$. Let π_x, π_y be an F-basis of M as left module, where π_x corresponds to a unirational point x (which exists by 0.6.2). That is, π_x is defined as the kernel of the universal extension

$$0 \longrightarrow L \xrightarrow{\pi_x} \sigma_x L \longrightarrow S_x \longrightarrow 0$$

where we identified M with $\mathrm{Hom}(L, \sigma_x L)$.

Let $\sigma = \sigma_x$ be the tubular shift associated to the unirational point x, and let $R = \Pi(L, \sigma)$ be the corresponding orbit algebra. Then π_x and π_y are elements in $R_1 = M$, and (at least) π_x is a prime element in R. (Note that π_y can be non-prime.) By Lemma 1.7.1 the element π_x is central.

By [89] there are $\alpha, \delta : F \longrightarrow F$ such that for all $f \in F$ the formula

$$\pi_y f = \delta(f)\pi_x + \alpha(f)\pi_y$$

holds, where α is a k-automorphism of F and δ is an $(\alpha, 1)$-derivation on F. Since $\dim_{R_0} R_n = n + 1$, it is easy to see that the $n + 1$ elements π_x^n, $\pi_x^{n-1}\pi_y$, $\pi_x^{n-2}\pi_y^2, \ldots, \pi_x\pi_y^{n-1}, \pi_y^n$ form a R_0-basis of R_n for each natural number n.

Denote by $F[X; Y, \alpha, \delta]$ the skew polynomial ring in two variables, where every element is expressible uniquely in the form $\sum_{i,j} f_{ij} X^i Y^j$ with $f_{ij} \in F$ (that is, as left polynomial) and such that X is central and for all $f \in F$ we have

$$Yf = \delta(f)X + \alpha(f)Y.$$

Since α is bijective each element can also be expressed uniquely as right polynomial. Moreover, this ring is graded by total degree (that is, $\deg(X) = 1 = \deg(Y)$). Then we get

PROPOSITION 1.7.3. *As graded algebras* $\Pi(L, \sigma_x) \simeq F[X; Y, \alpha, \delta]$. $\square$

Therefore this case is also referred to as the skew polynomial case.

REMARK 1.7.4. (1) See [89, 22] for an affine version of this. Therefore the preceding result is not surprising. One should expect that for simple bimodules (of any numerical type) one gets in a similar way graded analogues of the rings considered in [29, 23] and [18, 5.3].

(2) The function fields in the non-simple bimodule cases are well known, see [90]. From the preceding proposition it follows again that $k(\mathbb{X}) \simeq F(T, \alpha, \delta)$, the quotient division ring of the skew polynomial ring $F[T, \alpha, \delta]$. (The variable T is obtained as $T = YX^{-1}$.)

(3) The factoriality of the skew polynomial algebras in 1.7.3 also follows from results by Chatters and Jordan [13].

LEMMA 1.7.5. *Let* $R = F[X; Y, \alpha, \delta]$ *where* X *is central and* X *and* Y *having degree one. The homogeneous prime ideals of height one different from* RX *are in one-to-one correspondence with the non-zero prime ideals in* $F[Z, \alpha, \delta]$.

PROOF. Let R_X be the localization with respect to the central multiplicative set given by the powers X^n ($n \geq 0$). (See also Section 2.2.) By [36, 9.22] the homogeneous prime ideals disjoint from this set correspond to the homogeneous prime ideals in R_X. Restriction to the zero component gives the skew polynomial ring $F[Z, \alpha, \delta]$ in one variable, where $Z = YX^{-1}$. The prime ideals in this ring are in one-to-one correspondence with the homogeneous prime ideals in R_X since there is a central unit of degree one in R_X. $\qquad\square$

LEMMA 1.7.6. *In Proposition 1.7.3 one can assume either* $\alpha = 1_F$ *or* $\delta = 0$. *If* char $k = 0$ *then one can assume* $\delta = 0$.

PROOF. The first assertion follows by adopting [**44**, 1.1.21] to the case of two variables. If char $k = 0$ and $\Pi(L, \sigma_x) \simeq F[X; Y, 1_F, \delta]$, then one can adopt [**36**, 9.23] to show that δ is an inner derivation, and hence as graded algebras $\Pi(L, \sigma_x) \simeq F[X; Y, 1_F, 0]$. (If δ is not inner, then by [**36**, 9.23] the ring $F[Z, 1_F, \delta]$ is a simple ring. Then by 1.7.5, $\mathbb{X}$ consists of precisely one point. But since k is infinite there are infinitely many points by [**89**, Thm. 3]. We give an alternative argument: The zero component of the localization R_X is finitely generated over its centre ([**18**, 5.2]); on the other hand it coincides with $F[Z, 1_F, \delta]$, and in case δ is not inner this is not finitely generated over its centre by Amitsur's theorem (compare [**44**, 1.1.32]).) $\qquad\square$

If $\delta = 0$ then the non-simple bimodule case is also called the twisted polynomial case, if $\alpha = 1$ it is also called the differential polynomial case.

From now on we assume $\delta = 0$, so that $R = F[X; Y, \alpha]$ is a graded twisted polynomial algebra. Then also π_y is normal and hence prime. Let σ_y be the corresponding tubular shift. Modulo inner automorphisms α has finite order r. There is some $u \in F^*$ such that $\alpha^r(f) = u^{-1}fu$ for all $f \in F$. We have $M = F \oplus F$ with $f \cdot (x, y) = (fx, fy)$ and $(x, y) \cdot f = (xf, y\alpha(f))$ for all f, x, $y \in F$. We write also $M = M(F, \alpha)$.

Denote by $\mathrm{Fix}(\alpha)$ the subfield of all $f \in F$ such that $\alpha(f) = f$. With u as above, one can assume that $u \in \mathrm{Fix}(\alpha)^*$. Let $K = Z \cap \mathrm{Fix}(\alpha)$.

LEMMA 1.7.7. *The centre of* $R = F[X; Y, \alpha]$ *is given by* $K[X, uY^r]$. *The homogeneous prime elements in* R *are (up to multiplication with a unit)* X, Y *and the homogeneous prime elements in* $K[X, uY^r]$, *which are polynomials in* X^r *and* uY^r *with coefficients lying in* K.

PROOF. This follows by reducing to [**44**, 1.1.22] as in the proof of 1.7.5. $\qquad\square$

(Note that the centre can be also determined in the graded differential polynomial case $R = F[X; Y, 1_F, \delta]$, compare [**44**, 1.1.32].)

It follows in particular that (up to multiplication with a unit) every homogeneous prime element in R except Y is central. (Y itself is central (up to a unit) only in case $r = 1$.) Note that for example the central elements of the form $aX^r + buY^r$ with a, $b \in K^*$ are prime in R.

There is a version for the function field:

1.7.8. The centre of the function field $F(T, \alpha)$ is given by $K(uT^r)$, with the notations as above. The dimension of $F(T, \alpha)$ over its centre is given by $r^2 m^2$ with $m^2 = [F : Z(F)]$. (We call the number $s(\mathbb{X}) = rm$ the (global) skewness.)

LEMMA 1.7.9. *Let* z *be a point different from the points* x *and* y *which correspond to the prime elements* X *and* Y, *respectively. Then* $d(z) = e(z) \cdot f(z)$ *is a multiple of* r.

PROOF. The corresponding prime element π_z is a product of $e = e(z)$ many irreducible elements, each of degree $f = f(z)$. Now π_z is a homogeneous polynomial in the variables X^r and uY^r of degree $d = ef$. □

For an element $f \in F$ and each integer $i \geq 0$ let $N_i(f) = \alpha^{i-1}(f) \cdots \alpha(f)f$. We call $N(f) = N_r(f)$ the norm of f.

LEMMA 1.7.10. *With the notations as above, assume that there is an element* $b \in F$ *such that* $u^{-1} = N(b)$. *Then the element* $u^{-1}(X^r - uY^r)$ *is a product of* r *irreducible elements in* R. *Accordingly, the multiplicity of the corresponding rational point* z *is given by* $e(z) = r$, *and the endomorphism ring of the corresponding simple object* S_z *is given by the skew field of those elements* $f \in F$ *such that* $\alpha(f) = b^{-1}fb$.

PROOF. Follows from [**44**, 1.3.12] and 1.6.5. □

Explicit examples over the real numbers with complete lists of prime elements are given in 5.6.1.

1.7.11 (Arbitrarily large multiplicities). If F is commutative then one can assume that $u = 1$. If furthermore F/k is a cyclic Galois extension of degree r with Galois group generated by α then the preceding lemma can be applied. In this way it is possible to construct examples of exceptional curves having points (even rational ones) with arbitrarily large multiplicities.

The dimension of $k(\mathbb{X})$ over its centre is always a perfect square. This dimension can also be arbitrarily large which follows from the same example. In the present case $k(\mathbb{X}) = F(T, \alpha)$ has dimension r^2 over its centre $k(T^r)$.

It will follow from 2.2.13 that in this case r is the *maximal* multiplicity.

The quaternion case. Let k be a field of characteristic different from two. Let $a, b \in k^*$ and let $F = \left(\frac{a, b}{k} \right)$ be an algebra of quaternions over k, that is, a k-algebra on generators $\mathbf{i}$ and $\mathbf{j}$ subject to the relations

$$\mathbf{ji} = -\mathbf{ij}, \; \mathbf{i}^2 = a, \; \mathbf{j}^2 = b.$$

We assume that F is a skew field. Equivalently, the norm form of pure quaternions $-aX^2 - bY^2 + abZ^2$ is anisotropic over k. Let M be the bimodule ${}_kF_F$.

We have shown in [**54**] that the small preprojective algebra is given by

$$\Pi(L, \tau^-) \simeq k[X, Y, Z]/(-aX^2 - bY^2 + abZ^2).$$

We will see later that τ^- is the only efficient automorphism in this case. Note that the factoriality of this algebra was already known from a theorem of P. Samuel [**99**], we refer to [**33**, Prop. 11.5].

It is interesting that the bimodule ${}_kF_F$ given by noncommutative data gives a commutative orbit algebra. The next example shows the converse behaviour. A reason for this will be explained in Section 4.3.

The square roots case. The following example is based on calculations by D. Baer [**6**, 1.3.6]. In order to get the following result we need an additional argument.

PROPOSITION 1.7.12. *Let* $a, c \in k$ *and* K *the field* $k(\sqrt{a}, \sqrt{c})$ *such that* $[K : k] = 4$. *Let* M *be the* k-K-*bimodule* K *and let* $\mathbb{X}$ *be the associated homogeneous exceptional curve. Then there is a unirational point* $x \in \mathbb{X}$ *such that*

$$\Pi(L, \sigma_x) \simeq k\langle X, Y, Z \rangle/(XY - YX, XZ - ZX, YZ + ZY, Z^2 + aY^2 - cX^2),$$

where each variable is of degree one. In particular, the algebra on the right hand side is graded factorial.

PROOF. In [**6**, 1.3.6] the category of preprojective modules of rank one over the associated bimodule algebra, which is equivalent to the category $\mathcal{L}_+$ as defined in 1.4.1, is determined explicitly. It is easy to see that the morphism Y between two successive preprojectives of rank one described in [**6**] defines a universal extension with simple cokernel given as representation $S_x : k^2 \otimes K \xrightarrow{(1,\sqrt{a})} K$ (with endomorphism ring $k(\sqrt{a})$), and that the degree shift given in [**6**] coincides on the category $\mathcal{L}_+$ with the tubular shift σ_x. Now the assertion follows from the relations determined in [**6**] (where we changed the roles of the variables X and Y). $\square$

For further properties of this example see 4.3.7, 3.2.16 (2) and 5.7.2. Explicit examples are given by $k = \mathbb{Q}$ and $K = \mathbb{Q}(\sqrt{2}, \sqrt{3})$ or $K = \mathbb{Q}(\sqrt{2}, \sqrt{-3})$ (the latter case occurring in 8.3.2).

CHAPTER 2

Global and local structure of the sheaf category

As a consequence of a version of Serre's theorem by M. Artin and J. J. Zhang the graded factorial orbit algebras considered in the previous chapter are projective coordinate algebras for the homogeneous exceptional curves. It follows from the graded factoriality that homogeneous prime ideals of height one are localizable. We also consider the localization with respect to the multiplicative set given by the powers of a prime element. We describe properties of these localizations and derive important relations between the multiplicity function $x \mapsto e(x)$ and the dimension of the function field $k(\mathbb{X})$ over its centre.

2.1. Serre's theorem

Let $\mathbb{X}$ be a homogeneous exceptional curve with hereditary category $\mathcal{H}$. Let E be a non-zero object in $\mathcal{H}_+$, and let φ be some automorphism of $\mathcal{H}$. Recall that φ is positive if $\deg(\varphi L) > 0$. In [2] (see also [105]) the pair (E, φ) is called *ample* if the following holds:

(1) For each object $X \in \mathcal{H}$ there is an epimorphism $\bigoplus_{i=1}^{n} \varphi^{-\alpha_i}(E) \twoheadrightarrow X$ with integers $\alpha_i > 0$.
(2) Each epimorphism $X \twoheadrightarrow Y$ induces an epimorphism

$$\operatorname{Hom}(E, \varphi^n X) \twoheadrightarrow \operatorname{Hom}(E, \varphi^n Y)$$

for $n \gg 0$.

LEMMA 2.1.1. *If φ is positive then the pair (E, φ) is ample.*

PROOF. From the special structure of $\mathcal{H}$, the first property for ampleness is true for $E = L$ and $E = \overline{L}$ and then follows easily for arbitrary E. The second property for ampleness follows from 0.4.6 with Serre duality. $\square$

The following theorem is a special case of Serre's theorem for non-commutative projective schemes by M. Artin and J. J. Zhang [2].

PROPOSITION 2.1.2 (Serre's theorem). *Let $\mathbb{X}$ be a homogeneous exceptional curve. Let E be a non-zero vector bundle and φ be a positive automorphism of $\mathcal{H}$. Let $R = \Pi(E, \varphi) = \bigoplus_{n \geq 0} \operatorname{Hom}_{\mathcal{H}}(E, \varphi^n E)$ be the orbit algebra defined to the pair (E, φ). Then the section functors*

$$\Gamma : \mathcal{H} \longrightarrow \operatorname{Mod}^{\mathbb{Z}}(R), \ F \mapsto \bigoplus_{n \in \mathbb{Z}} \operatorname{Hom}(E, \varphi^n(F))$$

and

$$\Gamma_+ : \mathcal{H} \longrightarrow \operatorname{mod}^{\mathbb{Z}_+}(R), \ F \mapsto \bigoplus_{n \geq 0} \operatorname{Hom}(E, \varphi^n(F))$$

induce equivalences

$$\mathcal{H} \simeq \frac{\mathrm{mod}^{\mathbb{Z}}(R)}{\mathrm{mod}_0^{\mathbb{Z}}(R)} \simeq \frac{\mathrm{mod}^{\mathbb{Z}+}(R)}{\mathrm{mod}_0^{\mathbb{Z}+}(R)}.$$

$\square$

COROLLARY 2.1.3. *Each homogeneous exceptional curve admits a (noncommutative) graded factorial domain as projective coordinate algebra.* $\square$

COROLLARY 2.1.4. *Let $\mathbb{X}$ be an exceptional curve. Then the function field $k(\mathbb{X})$ is the quotient division ring of degree zero fractions of a graded factorial domain.*

PROOF. Note that the function field of an exceptional curve coincides with the function field of the underlying homogeneous curve. If $\mathbb{X}$ is homogeneous and $R = \Pi(L, \sigma)$ (with σ efficient) then by [**7**] (compare also [**88**, IV.4.1, Step 4]) $k(\mathbb{X})$ coincides with the degree zero part of the graded quotient division ring $\mathrm{Quot}^{\mathbb{Z}}(R)$ of R. $\square$

2.1.5 (The sheafification functor). Let $R = \Pi(E, \varphi)$ be with φ positive. (Of course, we have in mind $E = L$ and $\varphi = \sigma$ being efficient.) Denote by $T : \mathrm{mod}^{\mathbb{Z}}(R) \longrightarrow \frac{\mathrm{mod}^{\mathbb{Z}}(R)}{\mathrm{mod}_0^{\mathbb{Z}}(R)} \simeq \frac{\mathrm{mod}^{\mathbb{Z}+}(R)}{\mathrm{mod}_0^{\mathbb{Z}+}(R)}$ the canonical quotient functor. By the proposition TT_+ is an equivalence. Denote by $\phi : \frac{\mathrm{mod}^{\mathbb{Z}}(R)}{\mathrm{mod}_0^{\mathbb{Z}}(R)} \longrightarrow \mathcal{H}$ some quasi-inverse and define $\widetilde{\ } : \mathrm{mod}^{\mathbb{Z}}(R) \longrightarrow \mathcal{H}$ (and also its restriction to $\mathrm{mod}^{\mathbb{Z}+}(R)$) by $\widetilde{\ } = \phi T$. We may assume that $\widetilde{R(n)} = \varphi^n E$ (usually $= L(n)$) for all $n \in \mathbb{Z}$. This gives an exact and dense functor with kernel $\mathrm{mod}_0^{\mathbb{Z}}(R)$ and such that $\widetilde{\ } \circ \Gamma_+ \simeq 1$.

Denote also by T the quotient functor $T : \mathrm{Mod}^{\mathbb{Z}}(R) \longrightarrow \frac{\mathrm{Mod}^{\mathbb{Z}}(R)}{\mathrm{Mod}_0^{\mathbb{Z}}(R)}$. Then T admits a right adjoint S (the section functor) which is fully faithful and $TS = 1$ holds (compare [**85**, 4.4]). Denote by $S_+ : \frac{\mathrm{mod}^{\mathbb{Z}+}(R)}{\mathrm{mod}_0^{\mathbb{Z}+}(R)} \longrightarrow \mathrm{mod}^{\mathbb{Z}+}(R)$ the functor given by $S_+ T(M) = ST(M)_{\geq 0}$ (the non-negative part of $ST(M)$) for each $M \in \mathrm{mod}^{\mathbb{Z}+}(R)$. This is well-defined, since by [**2**, 4.5 S5-S7] $T(M) \simeq T\Gamma(F)$ for some $F \in \mathcal{H}$, and $ST(M)_{\geq 0} \simeq ST\Gamma(F)_{\geq 0} \simeq \Gamma(F)_{\geq 0} = \Gamma_+(F)$ is finitely generated. Since for all M and N in $\mathrm{mod}^{\mathbb{Z}+}(R)$ obviously $\mathrm{Hom}(M, ST(N)_{\geq 0}) = \mathrm{Hom}(M, ST(N))$, it follows directly by the adjointness of T and S that also (T, S_+) is an adjoint pair, that is, $\mathrm{mod}_0^{\mathbb{Z}+}(R)$ is a localizing subcategory of $\mathrm{mod}^{\mathbb{Z}+}(R)$. By the preceding argument, $S_+ T T_+ \simeq \Gamma_+$, and thus Γ_+ is fully faithful. Since there is a natural transformation $1 \longrightarrow S_+ T \simeq \Gamma_+ \circ \widetilde{\ }$, it follows easily that Γ_+ is right adjoint to $\widetilde{\ }$. $\square$

2.1.6 (Degree shift). We keep the notation from the preceding number. Denote by D the degree shift $X \mapsto X(1)$ on $\mathrm{Mod}^{\mathbb{Z}}(R)$. Then $D\Gamma = \Gamma\varphi$.

Denote by D_+ the functor on $\mathrm{mod}^{\mathbb{Z}+}(R)$ given by $D_+(M) = D(M)_{\geq 0}$. Then similarly $D_+ \Gamma_+ = \Gamma_+ \varphi$. It follows that via the equivalences TT_+ and ϕ the automorphism φ (defining the grading of R) corresponds to the automorphism on $\frac{\mathrm{mod}^{\mathbb{Z}+}(R)}{\mathrm{mod}_0^{\mathbb{Z}+}(R)}$ which is induced by D_+.

2.1.7 (Section modules/Cohen-Macaulay modules). For simplicity, we return to our standard situation, that is, $R = \Pi(L, \sigma)$ with σ being efficient. Then $R_0 = \mathrm{End}(L)$ is a skew field and is up to shift the only simple graded R-module. Denote by $\mathrm{sect}^{\mathbb{Z}+}(R)$ the full subcategory of $\mathrm{mod}^{\mathbb{Z}+}(R)$ formed by those M such that

$$\mathrm{Hom}_R(R_0(-n), M) = 0 = \mathrm{Ext}_R^1(R_0(-n), M)$$

for all $n \geq 0$. We have $M \in \mathrm{sect}^{\mathbb{Z}+}(R)$ if and only if there is $F \in \mathcal{H}$ such that $M \simeq \Gamma_+(F)$. In fact, from $M \simeq \Gamma_+(F)$ we get $\widetilde{M} \simeq F$. Moreover, it can be deduced from [**2**, 3.14] that $M \in \mathrm{sect}^{\mathbb{Z}+}(R)$ if and only if $M \simeq \Gamma_+(\widetilde{M})$.

Thus, the section functor Γ_+ induces an equivalence $\mathcal{H} \simeq \mathrm{sect}^{\mathbb{Z}+}(R)$.

Similarly, denote by $\mathrm{CM}^{\mathbb{Z}}(R)$ the full subcategory of $\mathrm{mod}^{\mathbb{Z}}(R)$ formed by those modules M such that

$$\mathrm{Hom}_R(R_0(n), M) = 0 = \mathrm{Ext}_R^1(R_0(n), M)$$

for all $n \in \mathbb{Z}$, that is, $\mathrm{CM}^{\mathbb{Z}}(R) = \mathrm{mod}_0^{\mathbb{Z}}(R)^{\perp}$. The objects in $\mathrm{CM}^{\mathbb{Z}}(R)$ are called graded maximal Cohen-Macaulay modules. Γ induces an equivalence $\mathcal{H}_+ \simeq \mathrm{CM}^{\mathbb{Z}}(R)$. For this, because of [**2**, 3.14] it is sufficient to show, that $\Gamma(F)$ is finitely generated for each $F \in \mathcal{H}_+$. Since by [**7**, 2.4] F is a subobject of a finite direct sum of shifts of L (and $\overline{L}$, if $\overline{L}$ is a line bundle, which then is a subobject of $L(1)$), this follows by left exactness of Γ and noetherianness of R. $\qquad\square$

2.1.8 (A Koszul complex). Let $R = \Pi(L, \sigma)$ with σ being an efficient automorphism. R has the Cohen-Macaulay property in the sense of regular sequences, since any two non-associated prime elements π_1 and π_2 define a regular sequence. For the (two-sided) ideal $I = \pi_1 R + \pi_2 R$ the factor R/I has Krull dimension zero, and hence is of finite length [**36**], and therefore finite dimensional.

Let d_1 and d_2 be the degrees of π_1 and π_2, respectively, and assume additionally that $\pi_1 \pi_2 = \pi_2 \pi_1$. Then this regular sequence defines a projective resolution of the graded right R-module R/I:

$$0 \longrightarrow R(-d_1 - d_2) \xrightarrow{\begin{pmatrix} \pi_2 \cdot \\ -\pi_1 \cdot \end{pmatrix}} R(-d_1) \oplus R(-d_2) \xrightarrow{(\pi_1 \cdot \ \pi_2 \cdot)} R \twoheadrightarrow R/I.$$

By sheafification this leads to the exact sequence

$$0 \longrightarrow L(-d_1 - d_2) \xrightarrow{\begin{pmatrix} \pi_2 \\ -\pi_1 \end{pmatrix}} L(-d_1) \oplus L(-d_2) \xrightarrow{(\pi_1 \ \pi_2)} L \longrightarrow 0.$$

$\qquad\square$

2.2. Localization at prime ideals

Let R be a graded ring and $S \subset R$ be a multiplicative set consisting of homogeneous elements. Following [**36**] we call S right Ore, if $rS \cap sR$ is non-empty for all $r \in R$ and $s \in S$. It is called a right reversible set, if for all $r \in R$ and $s \in S$ such that $sr = 0$ there is $s' \in S$ such that $rs' = 0$. (For both conditions it is sufficient to consider homogeneous elements r, see [**84**, I.6.1].) S is called a right denominator set if it is right Ore and right reversible. The left hand versions are defined similarly. S is called denominator set, if it is a left and right denominator set.

Let $\mathbb{X}$ be a homogeneous exceptional curve and $R = \Pi(L, \sigma)$ with σ being efficient. Since R is a graded domain, reversibility holds automatically for multiplicative sets. Moreover, each multiplicative set consisting of normal elements is a denominator set. For example, if $f \neq 0$ is normal, then $\{f^n \mid n \geq 0\}$ is a denominator set. Denote by $R_f = R[f^{-1}]$ the corresponding ring of fractions.

Let $P = R\pi$ be a prime ideal of height one and let $x \in \mathbb{X}$ be the corresponding point. We define the following multiplicative subsets of R consisting of homogeneous elements: $\mathcal{N}(P)$ contains all the normal elements not lying in P. The set Y consists of those non-zero homogeneous elements, where the induced map between line bundles has cokernel contained in $\coprod_{y \neq x} \mathcal{U}_y$, that is, is not supported by x. Let Y' be the set of all homogeneous elements s such that the fibre map s_x is an isomorphism. Finally, denote by $\mathcal{C}(P)$ the set of homogeneous elements in R whose classes are (left and right) regular (that is, non-zero divisors) in R/P.

LEMMA 2.2.1. *We have $\mathcal{N}(P) \subset Y = Y' = \mathcal{C}(P)$. All these multiplicative sets are denominator sets.*

PROOF. It is easy to see that all these sets are multiplicative and that

$$\mathcal{N}(P) \subset Y \subset Y' \subset \mathcal{C}(P).$$

$\mathcal{C}(P)$ consists of those (homogeneous) $s \in R$ such that $s_x \in \mathrm{M}_e(D)$ is regular (D a skew field), hence invertible (compare the proof of Theorem 1.2.3). Using Lemma 1.3.3, the equality $Y = Y'$ follows easily. Obviously, $\mathcal{N}(P)$ is a denominator set. That $\mathcal{C}(P)$ is a denominator set follows from a graded version of [**13**, Lemma 2.2]. In orbit case I and II we might use also 1.6.7 for the proof that Y' is a right Ore set: Let $r \in R$ be homogeneous and $s \in Y$. Without loss of generality let s be irreducible. There is some s_1 and a prime element π_1 such that $ss_1 = \pi_1$, where $R\pi_1 \neq R\pi$. Then

$$r\pi_1 = \pi_1 r' = s(s_1 r')$$

for some r', since π_1 is normal. The left hand version is similar. □

LEMMA 2.2.2. *Assume orbit case I or II. Then the graded rings of fractions $R_{\mathcal{N}(P)}$ and $R_{\mathcal{C}(P)}$ are isomorphic via the natural morphism.*

PROOF. (Compare also [**96**, 3.1.7].) Let $rs^{-1} \in R_{\mathcal{C}(P)}$. Compose $s = u_1 \ldots u_t$ into irreducibles. Then all $u_i \in \mathcal{C}(P)$. By 1.6.7 there are s_i and primes π_i such that $\pi_i = u_i s_i$. Then

$$rs^{-1} = ((rs_t)\pi_t^{-1})(s_{t-1}\pi_{t-1}^{-1}) \ldots (s_1\pi_1^{-1}) \in R_{\mathcal{N}(P)}.$$

Hence the natural morphism is surjective. Injectivity is proved along the same lines: If $rt^{-1} = 0$ in $R_{\mathcal{C}(P)}$ then there is some $s \in \mathcal{C}(P)$ with $rs = 0$. Decomposing s as above into irreducibles and representing irreducibles as divisors of primes as above we see that there is some $s' \in \mathcal{N}(P)$ such that $rs' = 0$, and hence $rt^{-1} = 0$ also in $R_{\mathcal{N}(P)}$. □

PROBLEM 2.2.3. Assume orbit case III. Does $R_{\mathcal{N}(P)} \simeq R_{\mathcal{C}(P)}$ also hold in this case? Is each irreducible element a (left and right) divisor of a normal element?

We denote the graded ring of fractions $R_{\mathcal{C}(P)}$ by R_P, its zero component by R_P^0. If P is corresponding to the point x, we also write R_x and R_x^0, respectively. We have

$$R_{\mathcal{N}(P)} = \bigcup_{\substack{f \notin P \\ f \text{ normal}}} R_f,$$

where R_f denotes the ring of fractions of the form rf^{-n}, with $r \in R$ and $n \geq 0$. Denote by $\mathrm{HOM}(F, G) = \bigoplus_{n \in \mathbb{Z}} \mathrm{Hom}(F, G(n))$ and $\mathrm{END}(F) = \mathrm{HOM}(F, F)$.

Denote by $\mathcal{H}_x$ the quotient category of $\mathcal{H}$ modulo the Serre subcategory spanned by $\coprod_{y \neq x} \mathcal{U}_y$. Denote by $q_x : \mathcal{H} \longrightarrow \mathcal{H}_x$, $F \mapsto F_x$ the canonical functor. We refer to [7] for more details about this category.

PROPOSITION 2.2.4. *Let L_x be the image of L in $\mathcal{H}_x$. Then*

$$R_P \simeq \mathrm{END}_{\mathcal{H}_x}(L_x).$$

PROOF. For homogeneous $as^{-1} \in R_P$, where a (resp. s) is of degree m (resp. n) in R, let $\rho(as^{-1}) = \sigma^{-n}(a) \circ (\sigma^{-n}(s))^{-1} \in \mathrm{Hom}_{\mathcal{H}_x}(L_x, L_x(m-n))$. It is easy to check that this is well-defined and gives an isomorphism of graded rings $\rho : R_P \longrightarrow \mathrm{END}_{\mathcal{H}_x}(L_x)$. $\qquad \square$

COROLLARY 2.2.5. *Let x, $y \in \mathbb{X}$ be points so that there is an automorphism $\varphi \in \mathrm{Aut}(\mathbb{X})$ with $\varphi(x) = y$. Then $R_x^0 \simeq R_y^0$.*

PROOF. The autoequivalence φ of $\mathcal{H}$ induces an equivalence $\mathcal{H}_x \xrightarrow{\sim} \mathcal{H}_y$ mapping L_x to L_y. $\qquad \square$

Recall that a (not necessarily commutative) ring is called a principal ideal domain if it is a domain and if every left ideal and every right ideal is generated by a single element.

PROPOSITION 2.2.6. $\mathcal{H}_x \simeq \mathrm{mod}^{\mathbb{Z}}(R_P) \simeq \mathrm{mod}(R_P^0)$. *In particular, R_P^0 is a principal ideal domain.*

PROOF. Note that R_P contains a unit u of degree one (take any non-zero element in R_1 with cokernel whose support is disjoint from x; the existence of such an element follows from [89, 3.6]). Therefore, the restriction to the zero component, $M \mapsto M_0$, induces an equivalence $\mathrm{mod}^{\mathbb{Z}}(R_P) \simeq \mathrm{mod}(R_P^0)$. By a similar argument, each line bundle in $\mathcal{H}$ becomes isomorphic to L_x in $\mathcal{H}_x$. Moreover, L_x is a projective generator of $\mathcal{H}_x$. It follows that $\mathcal{H}_x \simeq \mathrm{mod}(\mathrm{End}_{\mathcal{H}_x}(L_x))$, induced by the functor $\mathrm{Hom}_{\mathcal{H}_x}(L_x, -)$ (compare also [8, II.1.3]). By the result before, $\mathrm{End}_{\mathcal{H}_x}(L_x) \simeq R_P^0$.

As before, let $\Gamma(F) = \bigoplus_{n \in \mathbb{Z}} \mathrm{Hom}(L, F(n)) \in \mathrm{Mod}^{\mathbb{Z}}(R)$, and let $q_P : \mathrm{Mod}^{\mathbb{Z}}(R) \longrightarrow \mathrm{Mod}^{\mathbb{Z}}(R_P)$ be the canonical functor. It is easy to see that each element of $\mathrm{Ext}^1(L, F(n))$ can be annihilated by an element in R of sufficiently high degree (by 0.4.6), and that each morphism $L \longrightarrow S_y$ is annihilated by the element of R which is given by the kernel. It follows that the composition $q_p \circ \Gamma$ induces an exact functor $\mathcal{H} \longrightarrow \mathrm{mod}^{\mathbb{Z}}(R_P)$ such that for all $F \in \coprod_{y \neq x} \mathcal{U}_y$ we have $q_P \circ \Gamma(F) = 0$. Therefore we get the following diagram of functors:

$$
\begin{array}{ccc}
\mathcal{H} & & \\
\Big\downarrow{\scriptstyle q_x} & \searrow{\scriptstyle q_P \circ \Gamma} & \\
\mathcal{H}_x & \overset{\phi}{\dashrightarrow} \mathrm{mod}^{\mathbb{Z}}(R_P) & \overset{\simeq}{\longrightarrow} \mathrm{mod}(R_P^0)
\end{array}
$$

ϕ is the unique (exact) functor defined on the quotient category $\mathcal{H}_x$ such that $\phi \circ q_x = q_P \circ \Gamma$, and moreover, $\phi(F) = (\bigoplus_{n \in \mathbb{Z}} \mathrm{Hom}(L, F(n)))_P$ which is naturally isomorphic to $\bigoplus_{n \in \mathbb{Z}} \mathrm{Hom}(L_x, F_x(n))$ (similar to Proposition 2.2.4).

Each non-zero subobject of L_x is isomorphic to $L_x(-n)$ for some $n \geq 0$, hence R_P^0 is a right principal ideal domain. Since R_P^0 is noetherian it follows from [15] that R_P^0 is also a left principal ideal domain. $\qquad\square$

COROLLARY 2.2.7 (The structure of tubes). $\mathcal{U}_x \simeq \mathrm{mod}_0(R_P^0)$, *the modules of finite length over the principal ideal domain* R_P^0. $\qquad\square$

For a similar statement with a complete local domain occurring compare [90, Thm. 4.2] and 2.2.16 below.

PROPOSITION 2.2.8. *Let* $P = R\pi$ *be a homogeneous prime ideal of height one. There is a short exact sequence*

$$0 \longrightarrow R_P \stackrel{\pi\cdot}{\longrightarrow} R_P(d) \longrightarrow S_P^e \longrightarrow 0,$$

where S_P *is a simple graded right* R_P-*module and* e *the multiplicity of the corresponding point. In particular,* R_P/P_P *is a graded uniformly simple artinian ring.*

PROOF. Let $S \in \mathcal{H}$ be the simple object corresponding to P. Then $\phi \circ q_x(S)$ is a simple graded R_P-module, and applying the functor $\phi \circ q_x$ to the S-universal sequence given by π induces the short exact sequence as in the assertion. $\qquad\square$

For the notion of a (not necessarily commutative) Dedekind domain we refer to [77].

COROLLARY 2.2.9. R_P *is a graded Dedekind domain, that is, a noetherian hereditary domain such that each homogeneous, idempotent ideal equals* 0 *or* R_P. *In* R_P *the only homogeneous prime ideals are* 0 *and* P_P, *which is generated by a normal element. Moreover,* P_P *is the graded Jacobson radical of* R_P, *and each graded right torsion module is unfaithful. Each non-zero homogeneous ideal is a power of* P_P.

PROOF. Since R_P is a graded principal ideal domain, it follows that it is graded Dedekind. But there is also a direct argument: For each homogeneous element $0 \neq x \in R_P$ there is a natural number $n = v(x)$ such that $x \in P_P^n$, but $x \notin P_P^{n+1}$. If $I \neq 0$, R_P is an idempotent ideal, choose a non-zero homogeneous element $x \in I$ with $v(x)$ minimal in order to get a contradiction.

Moreover, the annihilator of each graded simple right module is given by $P_P \neq 0$, which follows by a graded version of [77, 4.3.18]. The last statement follows by a graded version of [77, 5.2.9]. $\qquad\square$

THEOREM 2.2.10. *Let* $P = R\pi$ *be a homogeneous prime ideal of height one and* S *be the associated simple sheaf. Let* D *be the endomorphism skew field* $\mathrm{End}(S)$ *and* e *the multiplicity of the corresponding point. Then* P_P^0 *is the unique non-zero prime ideal in* R_P^0 *and* $R_P^0/P_P^0 \simeq \mathrm{M}_e(D)$.

PROOF. In order to get the isomorphism apply the graded version of the Artin-Wedderburn theorem [84] and restrict to the zero component. Note that the endomorphism ring and the simplicity of S is preserved under the various functors we applied. It follows that P_P^0 is a maximal ideal in R_P^0, generated by the normal element $u^{-d}\pi 1^{-1}$ where $u \in R_P$ is a unit of degree one and d the degree of π. By the principal ideal theorem [77, 4.1.11] then P_P^0 is of height one.

Since P_P is the graded Jacobson radical of R_P, the zero component P_P^0 is the Jacobson radical of R_P^0. The Jacobson radical is the intersection of all primitive ideals. It thus follows that P_P^0 is the unique maximal ideal in R_P^0, and hence it is the only non-zero prime ideal in R_P^0.   □

REMARK 2.2.11. It follows in particular that R_P^0 and R_P are factorial (graded factorial, respectively).

DEFINITION 2.2.12. Let $s(\mathbb{X})$ be the square root of the dimension of $k(\mathbb{X})$ over its centre,
$$s(\mathbb{X}) = [k(\mathbb{X}) : Z(k(\mathbb{X}))]^{1/2}.$$
We call it the (global) *skewness* of $\mathbb{X}$. The curve $\mathbb{X}$ is commutative if and only if $s(\mathbb{X}) = 1$. For any point $x \in \mathbb{X}$ we call the square root $e^*(x)$ of the dimension of $\mathrm{End}(S_x)$ over its centre the *comultiplicity* of the point x. (These numbers are just the PI degrees of the respective skew fields.) The results below indicate that the multiplicity $e(x)$ plays the role of a "local skewness".

COROLLARY 2.2.13 (Upper bound for the multiplicities). *The multiplicities are bounded by the skewness $s(\mathbb{X})$.*
More precisely, with the comultiplicity $e^*(x)$ of a point x we have
$$e(x) \cdot e^*(x) \leq s(\mathbb{X}).$$

PROOF. This is now a direct consequence of general results on polynomial identities in the context of the Amitsur-Levitzki theorem, see [**77**, 13.3]. (The idea of the proof is due to W. Crawley-Boevey [**21**].) Let $e = e(x)$, $e^* = e^*(x)$, $s = s(\mathbb{X})$, $D = \mathrm{End}(S_x)$ and P be the corresponding prime ideal of height one. The localization R_P^0 is a subring of $k(\mathbb{X})$, and $\mathrm{M}_e(D)$ is a factor ring of R_P^0. Since $k(\mathbb{X})$ satisfies a (monic) polynomial identity of degree $2s$ (that is, s is the PI degree of $k(\mathbb{X})$) this holds also for $\mathrm{M}_e(D)$. But $2ee^*$ is the smallest degree of a polynomial identity for $\mathrm{M}_e(D)$, hence $ee^* \leq s$.   □

In particular:

COROLLARY 2.2.14. *If the function field $k(\mathbb{X})$ is commutative then $\mathbb{X}$ is multiplicity free and the endomorphism rings of the simple objects in $\mathcal{H}$ are commutative.*   □

In 4.3.1 it is shown that conversely the multiplicity freeness implies the commutativity.

The following simple observation is worth noting.

PROPOSITION 2.2.15. *If $e = 1$, then R_P^0 is local. If $e > 1$, then R_P^0 is not even semiperfect. The same is true for R_P in a graded version.*

PROOF. Since R_P^0 is a domain, 0 and 1 are the only idempotents in R_P^0. On the other hand, for $e > 1$ there are non-trivial idempotents in $R_P^0/P_P^0 \simeq \mathrm{M}_e(D)$.   □

2.2.16 (Completion). With the notations of 2.2.10, let $\widehat{R_P^0}$ be the P_P^0-adic completion of R_P^0. Then its Jacobson radical is given by $\widehat{P_P^0}$. By [**62**, 21.31+23.10] is $\widehat{R_P^0}$ a semiperfect ring with $\widehat{R_P^0}/\widehat{P_P^0} \simeq \mathrm{M}_e(D)$ and there is a complete local ring S_P such that $\widehat{R_P^0} \simeq \mathrm{M}_e(S_P)$, and $\mathcal{U}_x \simeq \mathrm{mod}_0(\widehat{R_P^0}) \simeq \mathrm{mod}_0(S_P)$ holds. Note that S_P does not longer contain information about the multiplicity e.

We have the following property which is well-known for commutative integrally closed noetherian domains [**76**, 11.5].

PROPOSITION 2.2.17.

$$R = \bigcap_{\mathrm{ht}(P)=1} R_{\mathcal{N}(P)}.$$

PROOF. If $s^{-1}r \in \cap R_{\mathcal{N}(P)}$, where s is normal, one has to show that $r \in sR$. By factorizing s in prime elements this follows easily. Compare also [**13**, Theorem 2.3]. □

2.3. Noncommutativity and the multiplicities

As in the previous section let $\mathbb{X}$ be homogeneous and $R = \Pi(L, \sigma)$ with σ efficient. Instead of localizing at a prime ideal, which means to "remove" all other points, we now exploit the localization with respect to elements whose cokernels are concentrated in a given point, which means to "remove" just this point. As application we get that the inequation in 2.2.13 is generically an equation.

LEMMA 2.3.1. *Let $x \in \mathbb{X}$ be a rational point. The set of non-zero homogeneous elements $s \in R$ such that $\mathrm{coker}(s) \in \mathcal{U}_x$ is a denominator set.*

PROOF. The right Ore condition: If $s : L \longrightarrow L(n)$ is in the defined set, and $r : L(m) \longrightarrow L(n)$ is a homogeneous element of R (up to shift), consider the inclusions i and j of the pullback (intersection) L' of s and r into L and $L(n)$, respectively. If L' is in the "wrong" orbit of line bundles, compose them with a monomorphism $f : L(p) \longrightarrow L'$ with cokernel in $\mathcal{U}_x$, which exists since x is rational. Then $sif = rjf$, and $\mathrm{coker}(jf) \in \mathcal{U}_x$.

For the left Ore condition we consider similarly a pushout diagram of r and s. If the obtained object of rank 1 decomposes, project to the line bundle summand. If this line bundle is in the "wrong" orbit, apply again a suitable map with cokernel in $\mathcal{U}_x$. (Compare [**7**, Lemma 2.6].) □

Let x be a rational point. We denote the localization of R with respect to the denominator set of the preceding lemma by $R_{\langle x \rangle}$, its degree zero component by $R_{\langle x \rangle}^0$. In this way we get as affine rings similar principal ideal domains like in [**18**, §5]:

PROPOSITION 2.3.2. *Let x be a rational point.*
(1) *As graded rings, $R_{\langle x \rangle} \simeq \mathrm{END}_{\mathcal{H}/\langle \mathcal{U}_x \rangle}(L)$.*
(2) *$\mathcal{H}/\langle \mathcal{U}_x \rangle \simeq \mathrm{mod}^{\mathbb{Z}}(R_{\langle x \rangle}) \simeq \mathrm{mod}(R_{\langle x \rangle}^0)$.*
(3) *$R_{\langle x \rangle}^0$ is a principal ideal domain.*

PROOF. Like in the preceding section. We only remark that there is a unit of degree one in $R_{\langle x \rangle}$. In fact, since x is rational, π can be written as $\pi = uv$ where u is irreducible and of degree one (compare 1.6.5 and 1.6.6), and then $(u \cdot 1^{-1}) \cdot (v \cdot \pi^{-1}) = 1 \cdot 1^{-1}$. □

LEMMA 2.3.3. *Let x be a rational point, $y \neq x$ and P_y the homogeneous prime ideal associated to y. Let r and s be homogeneous elements of R such that the cokernel of $s \neq 0$ lies in $\mathcal{U}_x$. Then $rs \in P_y$ implies $r \in P_y$.*

PROOF. The kernel and the cokernel of the fibre map s_y are on the one hand concentrated in x, on the other hand in y, hence they are zero and s_y is an isomorphism. By the proof of Theorem 1.2.3, $rs \in P_y$ means $\sigma^m(r_y) \circ s_y = (rs)_y = 0$ (where m is the degree of s), and thus $r_y = 0$ follows, which means $r \in P_y$. □

THEOREM 2.3.4. *Let x be a rational point. For each point y with $y \neq x$ denote by π_y' a normal generator of $\pi_y R_{\langle x \rangle} \cap R_{\langle x \rangle}^0$. Then*

(1) *We have $R_{\langle x \rangle}^0 / \pi_y' R_{\langle x \rangle}^0 \simeq \mathrm{M}_{e(y)}(D_y)$, where $D_y = \mathrm{End}(S_y)$.*

(2) *Let $U \subset \mathbb{X} \setminus \{x\}$ be an infinite subset. Then $\bigcap_{y \in U} \pi_y' R_{\langle x \rangle}^0 = 0$.*

(3) *There is a point $y \in U$ such that $e(y) \cdot e^*(y) = s(\mathbb{X})$.*

PROOF. (1) This follows as in 2.2.10 with a version of 2.2.8.

(2) Using the preceding lemma we get $\cap_{y \neq x} \pi_y R_{\langle x \rangle} = 0$ from 1.3.6, and intersecting with the component of degree zero gives the result.

(3) By (1) and (2) there is an inclusion $R_{\langle x \rangle}^0 \subset \prod_{y \in U} \mathrm{M}_{e(y)}(D_y)$. Since the PI degree of $R_{\langle x \rangle}^0$ coincides with the PI degree of its quotient division ring ([**1**, Thm. 7]), which is $k(\mathbb{X})$ (since there is a unit in $R_{\langle x \rangle}$ of degree one), the assertion follows. $\qquad \Box$

COROLLARY 2.3.5. *Let $\mathbb{X}$ be an exceptional curve. Then the equality*

$$e(y) \cdot e^*(y) = s(\mathbb{X})$$

holds generically, that is, for all points $y \in \mathbb{X}$ except finitely many.

PROOF. Let x be rational. The set $\{y \in \mathbb{X} \mid y \neq x, \ e(y) \cdot e^*(y) \neq s(\mathbb{X})\}$ must be finite by part (3) of the preceding theorem. (Obviously, this holds also in the weighted situation.) $\qquad \Box$

REMARK 2.3.6. In general points x such that $e(x) \cdot e^*(x) \neq s(\mathbb{X})$ may exist. For example, for the bimodule $M = \mathbb{C} \oplus \overline{\mathbb{C}}$ over the real numbers this inequality holds precisely for the points corresponding to the prime ideals generated by X and Y, respectively, in the orbit algebra $\mathbb{C}[X; Y, \bar{\cdot}]$. (Note that these two points are just the separation points on the boundary. See 5.6.1 (5) for more details.)

It is an interesting question which role those finitely many points play where inequality holds and whether there is a connection to the ghost group.

COROLLARY 2.3.7. *Assume that there is an infinite subset U of points x whose corresponding simple sheaves S_x all have commutative endomorphism rings. Then there is a point $x \in U$ such that $e(x) = s(\mathbb{X})$.* $\qquad \Box$

COROLLARY 2.3.8. *Let $\mathbb{X}$ be an exceptional curve over a* finite *field k. Then there is a point $x \in \mathbb{X}$ such that $e(x) = s(\mathbb{X})$. Actually, this holds for all except finitely many points.*

PROOF. Since each finite skew field is commutative (Wedderburn), we have $e^*(x) = 1$ for all $x \in \mathbb{X}$. Since $\mathbb{X}$ contains always infinitely many points, the result follows from the preceding results. $\qquad \Box$

COROLLARY 2.3.9. *Let k be a finite field. Then $\mathbb{X}$ is commutative if and only if $\mathbb{X}$ is multiplicity free.*

PROOF. Follows directly from 2.2.13 and 2.3.8. $\qquad \Box$

We will show in 4.3.1 that the preceding corollary holds over any field. Hence our results show that the multiplicity function e measures noncommutativity, locally and globally.

PROBLEM 2.3.10. Is it true that $s(\mathbb{X})$ is always the maximum of e? Is this maximum taken on even by a rational point? Are there always infinitely many points (even rational points, if k is infinite) x such that $\mathrm{End}(S_x)$ is commutative? (For the class of examples in 1.7.11 the answer to all three questions is positive.)

PROBLEM 2.3.11. Understand the role each single multiplicity $e(x)$ (and co-multiplicity $e^*(x)$) plays in terms of the function field. In particular: Is each $e(x)$ and $e^*(x)$ or their product a divisor of $s(\mathbb{X})$?

2.4. Localizing with respect to the powers of a prime element

In this section we describe affine parts of the curve $\mathbb{X}$ by localizing with respect to the powers of certain prime elements.

LEMMA 2.4.1. *Let $x \in \mathbb{X}$, and assume that $f(x)$ is even in orbit case III. Let $\pi_x \in R$ be the corresponding prime element and S_x the corresponding simple object.*
(1) Let $s \in R$ be homogeneous such that $s \neq 0$ and $\mathrm{coker}(s) \in \mathcal{U}_x$. Then there is factorization $s = u_1 \ldots u_t$, where each $u_i \in R$ is irreducible such that $\mathrm{coker}(u_i) \simeq S_x$. Moreover, each u_i is a (left and right) divisor of π_x.
(2) The multiplicative subset of all homogeneous $s \in R$ such that $s \neq 0$ and $\mathrm{coker}(s) \in \mathcal{U}_x$ is a denominator set.
(3) Each fraction rs^{-1}, with $s \neq 0$ homogeneous such that $\mathrm{coker}(s) \in \mathcal{U}_x$, can be written as $r'\pi_x^{-n}$ for some $r' \in R$ and some $n \geq 0$.

PROOF. (1) The factorization of s follows as in the proof of Proposition 1.6.6 considering a chain of projections from the cokernel of s decreasing the length in each step by one; the cokernel of each u_i is isomorphic to S_x (since in orbit case III we assume that $f(x)$ is even). Each u_i is a divisor of π_x by Lemma 1.6.7.
(2) Follows as in the proof of Lemma 2.2.1.
(3) Follows by a similar argument given in the proof of Lemma 2.2.2. □

2.4.2 (Quasi-rational points). In orbit case III, ring theoretically rational points behave more complicated in some sense than points x with $f(x) = 2$, compare Proposition 1.6.6, or the preceding lemma.

We call a point $x \in \mathbb{X}$ *quasi-rational* if the prime element π_x factors into irreducibles of degree one and with simple cokernel. That is,

$$x \text{ is quasi-rational} \iff \begin{cases} f(x) = 1 & \text{in orbit cases I and II,} \\ f(x) = 2 & \text{in orbit case III.} \end{cases}$$

Since rational points always exist, trivially also quasi-rational points exist in orbit cases I and II. In orbit case III, a quasi-rational point exists if and only if there is a non-zero map $f \in \mathrm{Hom}(L, L(1))$ which cannot be decomposed into a product of maps from $\mathrm{Hom}(L, \overline{L})$ and $\mathrm{Hom}(\overline{L}, L(1))$, that is, f is 1-irreducible. The existence of such a map is open in general, but one should expect that this follows by a similar dimension argument like in [**29**, 2.4].

PROBLEM 2.4.3. Do quasi-rational points always (in orbit case III) exist?

PROPOSITION 2.4.4. *Let $\pi = \pi_x \in R$ be a prime element corresponding to a quasi-rational point x. Let $R_\pi = R[\pi^{-1}] = \{r\pi^{-n} \mid r \in R, n \geq 0\}$. Denote by R_π^0 its component of elements of degree zero.*
(1) As graded rings, $R_\pi \simeq \mathrm{END}_{\mathcal{H}/\langle \mathcal{U}_x \rangle}(L)$.

(2) $\mathcal{H}/\langle \mathcal{U}_x \rangle \simeq \mathrm{mod}^{\mathbb{Z}}(R_\pi) \simeq \mathrm{mod}(R_\pi^0)$.

(3) R_π^0 *is a principal ideal domain.*

PROOF. (1) Note that each homogeneous fraction in $\mathrm{END}_{\mathcal{H}/\langle \mathcal{U}_x \rangle}(L)$ can be written as fraction rs^{-1} such that r, $s \in R$ are homogeneous with $s \neq 0$ and $\mathrm{coker}(s) \in \mathcal{U}_x$. With the preceding lemma the assertion follows as in Proposition 2.2.4.

(2) and (3) follow similarly as in the proof of Proposition 2.2.6. Note that there is a unit of degree one in R_π. In fact, since x is quasi-rational, π can be written as $\pi = uv$ where u is irreducible and of degree one (compare 1.6.5 and 1.6.6), and then $(u \cdot 1^{-1}) \cdot (v \cdot \pi^{-1}) = 1 \cdot 1^{-1}$. $\qquad \square$

2.5. Zariski topology and sheafification

As before, let $\mathbb{X}$ be homogeneous with hereditary category $\mathcal{H}$ and orbit algebra $R = \Pi(L, \sigma)$, where σ is efficient.

From the Artin-Zhang version of Serre's theorem we deduced some formal "sheafification functor" $\widetilde{\ } : \mathrm{mod}^{\mathbb{Z}}(R) \longrightarrow \mathcal{H}$. In this section we briefly sketch how we get such a functor in a more explicit manner.

We identify $\mathbb{X}$ with the set of height one homogeneous prime ideals in R, hence $\mathbb{X} = \mathrm{Proj}(R)$. If f is a normal element, denote by $\mathbb{X}_f$ the subset of all prime ideals P such that $f \notin P$. The following lemma is easy to show.

LEMMA 2.5.1. *The system of sets $\mathbb{X}_f$ ($f \neq 0$ normal, non-unit) forms a basis for a topology such that $\mathbb{X}$ is connected and the proper closed sets are precisely the finite subsets of $\mathbb{X}$.* $\qquad \square$

It is routine to define graded coherent and quasi-coherent sheaves over $\mathbb{X}$ and we get a sheafification functor $\widetilde{\ } : \mathrm{Mod}^{\mathbb{Z}}(R) \longrightarrow \mathrm{Qcoh}(\mathbb{X})$, $M \mapsto \widetilde{M}$, where $\widetilde{M}(\mathbb{X}_f) = M_f$, the localization of M with respect to the powers of f, where the structure sheaf $\mathcal{O}_{\mathbb{X}}$ is defined by $\mathcal{O}_{\mathbb{X}}(\mathbb{X}_f) = R_f$. For each point $x \in \mathbb{X}$ the stalk is given by the localization $R_{\mathcal{N}(P)}$, where P is the homogeneous prime ideal corresponding to x. By 2.2.17 it follows that $\mathcal{O}_{\mathbb{X}}(\mathbb{X}) = R$. For the global section functor $\Gamma = \Gamma(\mathbb{X}, -) : \mathrm{Qcoh}(\mathbb{X}) \longrightarrow \mathrm{Mod}^{\mathbb{Z}}(R)$ one shows the following properties:

- Γ is right adjoint to $\widetilde{\ }$ (using the natural morphism $\rho_M : M \longrightarrow \widetilde{M}(\mathbb{X})$, assigning to $m \in M$ all the fractions $m1^{-1} \in M_f = \widetilde{M}(\mathbb{X}_f)$).
- $\widetilde{\ } \circ \Gamma \simeq 1$.

Moreover, there is the following version of Serre's theorem.

PROPOSITION 2.5.2 (Serre's theorem). *Sheafification*

$$\widetilde{\ } : \mathrm{mod}^{\mathbb{Z}}(R) \longrightarrow \mathrm{coh}(\mathbb{X})$$

induces an equivalence

$$\mathrm{coh}(\mathbb{X}) \simeq \frac{\mathrm{mod}^{\mathbb{Z}}(R)}{\mathrm{mod}_0^{\mathbb{Z}}(R)}.$$

PROOF. We only show that $\mathrm{mod}_0^{\mathbb{Z}}(R)$ is the kernel of the sheafification functor. Denote by $\mathfrak{m}$ the graded Jacobson radical of R. We have $\widetilde{M} = 0$ if and only if $M_f = 0$ for all normal elements $f \in \mathfrak{m}$. Since each graded simple R-module is of the form $(R/\mathfrak{m})(n)$ for some $n \in \mathbb{Z}$, we see that $\widetilde{M} = 0$ if M is of finite length.

For $M \in \mathrm{mod}^{\mathbb{Z}}(R)$ there is a finite sequence $0 = M_0 \subset \cdots \subset M_s = M$ of submodules such that M_i/M_{i-1} is a fully faithful (and torsionfree) left R/P_i-module, where P_i is the corresponding affiliated prime ideal (see [**36**, 2.+8.]). If $\widetilde{M} = 0$, then for all normal $f \in \mathfrak{m}$ it follows that f^n is contained in the annihilator of M for some $n \geq 0$, hence $f \in P_i$ and thus $P_i = \mathfrak{m}$ for all $i = 1, \ldots, s$. Therefore M is of finite length.   □

COROLLARY 2.5.3. $\mathcal{H}$ *is equivalent to* $\mathrm{coh}(\mathbb{X})$.   □

There is the commutative diagram

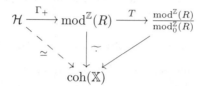

where T is the quotient functor and $T \circ \Gamma_+$ gives an equivalence. Identifying $\mathcal{H}$ with $\mathrm{coh}(\mathbb{X})$ via the equivalence $\widetilde{\ } \circ \Gamma_+$, it is clear that the efficient automorphism σ (defining R) corresponds to the degree shift on $\mathrm{mod}^{\mathbb{Z}}(R)$, $X \mapsto X(1)$. Moreover, $\widetilde{R(n)} = L(n)$.

We get the following statement which is a generalization of the statement which sometimes is referred to as Grothendieck's lemma if $\mathbb{X} = \mathbb{P}^1(\overline{k})$.

COROLLARY 2.5.4. *Each indecomposable vector bundle over* $\mathbb{X}$ *is isomorphic to* $L(n)$ *or to* $\overline{L}(n)$ *for some* $n \in \mathbb{Z}$.   □

Here, the number n is unique in orbit cases I and III; in orbit case II one has $\overline{L} = L(1)$. Moreover, $\overline{L}$ is not a line bundle precisely in orbit case I.

CHAPTER 3

Tubular shifts and prime elements

This chapter is motivated by the question when two tubular shifts σ_x and σ_y coincide. Whereas the answer over an algebraically closed field is trivial this question becomes interesting over an arbitrary field and we will show that it is linked to (the centre of) the function field.

In Chapter 1 we got the graded factoriality of $R = \Pi(L, \sigma)$ (with σ efficient) by showing that there is a one-to-one correspondence between points $x \in \mathbb{X}$ and prime elements π_x (up to multiplication with some unit), naturally given by forming universal extensions. In this chapter we show that this actually extends to a natural correspondence between prime elements π_x and tubular shifts σ_x together with the natural transformations $1_{\mathcal{H}} \xrightarrow{x} \sigma_x$; in fact, such a transformation is induced by right multiplication with π_x.

As a consequence we get a relation between the tubular shifts and the degree shift. The difference between tubular shifts and the degree shift is given by ghosts arising from prime elements which are not central (up to multiplication with a unit). Thus we get important information on the structure of the Picard group. In particular, this group is not always torsionfree.

3.1. Central prime elements

Let $\mathbb{X}$ be a homogeneous exceptional curve with structure sheaf L and σ be an efficient automorphism. Let $R = \Pi(L, \sigma)$ be the corresponding orbit algebra. Recall, that we denote $\sigma^n(F)$ also by $F(n)$ for $F \in \mathcal{H}$ and also for $F \in \mathrm{mod}^{\mathbb{Z}}(R)$ (degree shift).

3.1.1 (Central multiplication). It is convenient to consider first the special, central case. Let $r \in R$ be a central homogeneous element of degree n. Then multiplication from the right with r induces a homomorphism $M \xrightarrow{\cdot r} M(n)$ for each $M \in \mathrm{mod}^{\mathbb{Z}}(R)$ (right modules), and by sheafification, this induces also a morphism $\widetilde{M} \xrightarrow{\cdot r} \widetilde{M}(n)$ in $\mathcal{H}$. We obtain a natural transformation $1_{\mathcal{H}} \xrightarrow{\cdot r} \sigma^n$. It follows easily that

- for $M = R$ we get $\cdot r = r \in \mathrm{Hom}(L, L(n))$;
- if we apply the functor Γ_+ to $F \xrightarrow{\cdot r} F(n)$ in $\mathcal{H}$, then we get back multiplication with r on the level of graded right R-modules.

THEOREM 3.1.2. *Let $R = \Pi(L, \sigma)$ with σ being efficient. Let $x \in \mathbb{X}$ such that the prime element $\pi_x \in R$ is central of degree d. Then there is a natural isomorphism from the tubular shift σ_x to the degree shift σ^d, which is compatible with the natural transformations $1_{\mathcal{H}} \xrightarrow{x} \sigma_x$ and $1_{\mathcal{H}} \xrightarrow{\cdot \pi_x} \sigma^d$.*

PROOF. Let $M \in \mathcal{H}_+$ be of rank $r > 0$. Let $S = S_x$ be the simple object concentrated in x. There is the universal extension

$$0 \longrightarrow M \xrightarrow{\alpha_M} \sigma_x(M) \longrightarrow \mathrm{Ext}^1(S, M) \otimes_{\mathrm{End}(S)} S \longrightarrow 0.$$

The morphism $M \xrightarrow{\cdot \pi_x} M(d)$ is a monomorphism and its cokernel is concentrated in the point x and of length $r \cdot e(x)$: using a line bundle filtration of M and induction, it suffices to show this for line bundles L'. If L' is a shift of L then the cokernel is $S_x^{e(x)}$. In orbit case III we have to consider also the case $L' = \overline{L}$. There is an irreducible map $L \xrightarrow{u} \overline{L}$, and the cokernel S_0 is a simple object. One can assume that $S_0 \not\simeq S_x$. It then follows that the cokernel of the map $\overline{L} \xrightarrow{\cdot \pi_x} \overline{L}(d)$ is also isomorphic to $S_x^{e(x)}$.

We have to show that the cokernel C of $M \xrightarrow{\cdot \pi_x} M(d)$ is semisimple: The map $C \xrightarrow{\cdot \pi_x} C(d)$ is zero, which follows from the commutative exact diagram

$$
\begin{array}{ccccccccc}
0 & \longrightarrow & M & \xrightarrow{\cdot \pi_x} & M(d) & \xrightarrow{p} & C & \longrightarrow & 0 \\
& & \downarrow{\scriptstyle \cdot \pi_x} & & \downarrow{\scriptstyle \cdot \pi_x(d)} & & \downarrow{\scriptstyle \cdot \pi_x} & & \\
0 & \longrightarrow & M(d) & \xrightarrow{\cdot \pi_x(d)} & M(2d) & \xrightarrow{p(d)} & C(d) & \longrightarrow & 0.
\end{array}
$$

Applying the exact functor $\phi \circ q_x$ from 2.2.6 we get a short exact sequence of right R_P-modules:

$$0 \longrightarrow M_P \xrightarrow{\cdot \pi_x} M(d)_P \longrightarrow C_P \longrightarrow 0.$$

Since the map $C \xrightarrow{\cdot \pi_x} C(d)$ is zero, we see that C_P is a graded R_P/P_P-module, hence semisimple by 2.2.8. It follows that C is a direct sum of copies of S.

Hence we get a commutative, exact diagram

$$
\begin{array}{ccccccccc}
0 & \longrightarrow & M & \xrightarrow{\alpha_M} & M(x) & \xrightarrow{\beta_M} & M_x & \longrightarrow & 0 \\
& & \| & & \uparrow{\scriptstyle i_M} & & \uparrow{\scriptstyle j_M} & & \\
0 & \longrightarrow & M & \xrightarrow{\cdot \pi_x} & M(d) & \longrightarrow & C & \longrightarrow & 0,
\end{array}
$$

with isomorphisms i_M and j_M (compare 0.4.2 (2)). With the uniqueness property 0.4.2 (1) of $\sigma_x(f)$ it follows easily that for each $f \in \mathrm{Hom}(M, N)$ (with $N \in \mathcal{H}_+$) we have $\sigma_x(f) \circ i_M = i_N \circ f(d)$. Therefore, the functors σ^d and σ_x are naturally isomorphic on $\mathcal{H}_+$. (Actually, the argument shows, that this holds everywhere outside $\mathcal{U}_x$.) With Lemma 1.2.2, presenting each object in $\mathcal{H}_0$ as cokernel of a monomorphism between objects from $\mathcal{H}_+$, the result follows by diagram chasing. $\square$

COROLLARY 3.1.3. *Let x, $y \in \mathbb{X}$ such that the corresponding prime elements π_x and π_y are central of the same degree in R. Then the tubular shifts σ_x and σ_y are isomorphic.* $\square$

3.2. Non-central prime elements and ghosts

3.2.1 (Normal multiplication). We generalize Theorem 3.1.2 to arbitrary prime elements. Therefore, assume that $r \in R$ is a (non-zero) normal element of degree n. Then r induces the automorphism $\gamma = \gamma_r$ of the graded algebra R, by the formula

$sr = r\gamma(s)$. Let $M \in \mathrm{mod}^{\mathbb{Z}}(R)$. Denote by M_γ the object in $\mathrm{mod}^{\mathbb{Z}}(R)$, where M and M_γ coincide as abelian groups, and where the R-action on M_γ is defined by

$$m \cdot_\gamma s \overset{def}{=} m \cdot \gamma(s).$$

Then multiplication $\cdot r : m \mapsto m \cdot r$ defines a morphism between the graded right modules M and $M_\gamma(n)$, since

$$(m \cdot s) \cdot r = m \cdot (s \cdot r) = m \cdot (r \cdot \gamma(s)) = (m \cdot r) \cdot_\gamma s.$$

Since obviously $\gamma(\mathfrak{m}) = \mathfrak{m}$, the algebra automorphism γ gives rise to an automorphism γ^* by $\gamma^*(M) = M_\gamma$ on $\mathrm{mod}^{\mathbb{Z}}(R)$ and (denoted by the same symbol) also on $\mathcal{H} = \frac{\mathrm{mod}^{\mathbb{Z}}(R)}{\mathrm{mod}_0^{\mathbb{Z}}(R)}$, such that $\sigma \circ \gamma^* = \gamma^* \circ \sigma$; moreover, $\gamma^*(\widetilde{M}) = \widetilde{M_\gamma}$. Sheafification gives a homomorphism $\widetilde{M} \xrightarrow{\cdot r} \widetilde{M_\gamma}(n)$, yielding a natural transformation $1_{\mathcal{H}} \xrightarrow{\cdot r} \sigma^n \circ \gamma^* = \gamma^* \circ \sigma^n$. Obviously, γ gives rise to an isomorphism $R \longrightarrow R_\gamma$ of graded right R-modules, hence γ^* leaves L fixed, that is, $\gamma^* \in \mathrm{Aut}(\mathbb{X})$. Note that after identifying R with R_γ via γ, the functor γ^* on $\mathrm{mod}^{\mathbb{Z}}(R)$ acts on elements $f \cdot$ of $\mathrm{Hom}(R, R(m))$ (in contrast to $\mathrm{Hom}(R_\gamma, R_\gamma(m))$!) like γ^{-1}, since $(\gamma^{-1} \circ (\gamma^*(f\cdot)) \circ \gamma)(r) = \gamma^{-1}(f) \cdot r$. We will also make use of the notation $\gamma_* = (\gamma^{-1})^*$. □

The k-algebra automorphisms induced by normal elements are special cases of a more general class of graded algebra automorphisms:

DEFINITION 3.2.2. An automorphism γ of the graded k-algebra R is called *prime fixing*, if for all homogeneous prime ideals P (of height one) we have $\gamma(P) = P$. In other words, γ is prime fixing if and only if for each prime element π there is a unit $u \in R_0^*$ such that $\gamma(\pi) = \pi u$. Denote by $\mathrm{Aut}_0(R)$ the subgroup of $\mathrm{Aut}(R)$ of all *graded* algebra automorphisms, which are prime fixing.

Examples of prime fixing automorphism are the inner automorphisms ι_u ($u \in R_0^*$) given by $\iota_u(r) = u^{-1}ru$, defining the subgroup $\mathrm{Inn}(R)$, and automorphisms φ_a defined in the following way: Let $a = (a_i)_{i\geq 1}$ be a sequence of elements $a_i \in Z(R_0)^*$ with $ra_i = a_{i+1}r$ for all homogeneous elements $r \in R$ of degree one. Then $\varphi_a(r) = a_1 \cdot a_2 \cdots a_n \cdot r$ (for each homogeneous $r \in R$ of degree n) defines an automorphism $\varphi_a \in \mathrm{Aut}_0(R)$. (Recall that R is generated in degrees 0 and 1.) We denote the subgroup of $\mathrm{Aut}(R)$ generated by all ι_u and all φ_a by $\overline{\mathrm{Inn}}(R)$ which is a normal subgroup of $\mathrm{Aut}(R)$. □

Let $a = (a_i)$ be a sequence as above defining φ_a. Since R contains central homogeneous elements (see 4.1.3) there is $n \geq 1$ such that

(3.2.1) $$a_{i+n} = a_i \quad \text{for all } i \geq 1.$$

If $ra_1 = a_1 r$ for some r of degree one then $n = 1$ can be chosen, and a_1 lies in the centre of R. This is the case, for example, if there is a central element of degree one in R.

PROPOSITION 3.2.3. (1) *Let γ be a graded algebra automorphism of R. Then the induced automorphisms γ_* and γ^* on $\mathcal{H}$ are automorphisms of $\mathbb{X}$. Moreover, γ^* is trivial (that is, isomorphic to the identity) if and only if $\gamma \in \overline{\mathrm{Inn}}(R)$. Hence the assignment $\gamma \mapsto \gamma_*$ induces an injective group homomorphism from $\mathrm{Aut}(R)/\overline{\mathrm{Inn}}(R)$ into $\mathrm{Aut}(\mathbb{X})$.*

(2) *Let γ be a prime fixing automorphism. Then the induced automorphisms γ_* and γ^* on $\mathcal{H}$ are elements of the ghost group. Hence, there is an injective group homomorphism from $\mathrm{Aut}_0(R)/\overline{\mathrm{Inn}}(R)$ into the ghost group $\mathcal{G}$.*

(3) *Let $r \in R$ be normal. The associated algebra automorphism $\gamma = \gamma_r$ is prime fixing, and it is inner if and only if there is some unit $u \in R_0^*$ such that ru is central.*

PROOF. (1) As already seen above γ^* fixes L, hence is an automorphism of $\mathbb{X}$. Using [**81**, IV.5] the automorphism γ induces the identity on $\mathrm{Mod}^{\mathbb{Z}}(R)$ (or $\mathrm{mod}^{\mathbb{Z}}(R)$) if and only if the restriction of γ^* to the full subcategory generated by the shifts $R(n)$ ($n \in \mathbb{Z}$) is isomorphic to the identity, that is, for each integer i there are commutative diagrams

$$(3.2.2) \qquad \begin{array}{ccc} R(i-1) & \xrightarrow[\sim]{f_{i-1}} & R_\gamma(i-1) \\ {\scriptstyle r\cdot}\downarrow & & \downarrow{\scriptstyle r\cdot} \\ R(i) & \xrightarrow[\sim]{f_i} & R_\gamma(i) \end{array}$$

for all $r \in R$ of degree one (and similarly of degree zero). If this is the case, setting $u = f_0(1)$, $a_i f_i(1) = f_{i-1}(1)$ for all $i \geq 1$, then $\gamma = \iota_u \circ \varphi_a$ follows. For the converse, if $\gamma = \iota_u$ is inner, then $f_i(s) = u\gamma(s)$ defines isomorphisms making (3.2.2) commutative. If $\gamma = \varphi_a$ is given by a sequence $a = (a_i)_{i \geq 1}$ as above then $f_i(s) = a_i^{-1} \cdot \ldots \cdot a_1^{-1} \cdot \gamma(s)$ defines isomorphisms for $i \geq 0$ such that the diagrams (3.2.2) commute for each $i \geq 1$. With (3.2.1) isomorphisms f_i can be defined also for $i < 0$ in a similar way so that for each integer i the diagrams (3.2.2) commute.

If γ induces the identity on $\mathrm{mod}^{\mathbb{Z}}(R)$ then also on the quotient category $\mathcal{H} = \mathrm{mod}^{\mathbb{Z}}(R)/\mathrm{mod}_0^{\mathbb{Z}}(R)$. Using the section functor Γ the converse follows with 2.1.7.

(2) Let $x \in \mathbb{X}$ and π_x be the associated prime element of degree d. By assumption, there is some $u \in R_0^*$ such that $\gamma(\pi_x) = \pi_x u$. Then $L \xrightarrow{\pi_x} L(d)$ induces (via Γ_+) the commutative, exact diagram of right graded R-modules

$$\begin{array}{ccccccccc} 0 & \longrightarrow & R & \xrightarrow{\pi_x\cdot} & R(d) & \longrightarrow & R/\pi_x R & \longrightarrow & 0 \\ & & {\scriptstyle (u\cdot)\circ\gamma}\downarrow{\scriptstyle \simeq} & & {\scriptstyle \simeq}\downarrow{\scriptstyle \gamma(d)} & & \downarrow & & \\ 0 & \longrightarrow & R_\gamma & \xrightarrow{\pi_x\cdot} & R_\gamma(d) & \longrightarrow & R_\gamma/\pi_x R_\gamma & \longrightarrow & 0, \end{array}$$

and sheafification implies (using $\widetilde{} \circ \Gamma_+ = 1_{\mathcal{H}}$)

$$(3.2.3) \qquad \begin{array}{ccccccccc} 0 & \longrightarrow & L & \xrightarrow{\pi_x} & L(d) & \longrightarrow & L_x & \longrightarrow & 0 \\ & & {\scriptstyle \simeq}\downarrow & & \downarrow{\scriptstyle \simeq} & & \downarrow{\scriptstyle \simeq} & & \\ 0 & \longrightarrow & \gamma^*(L) & \xrightarrow{\pi_x\cdot} & \gamma^*(L)(d) & \longrightarrow & \gamma^*(L_x) & \longrightarrow & 0. \end{array}$$

Since γ^* fixes $L_x = S_x^{e(x)}$, the point x is fixed. It follows that $\gamma^* \in \mathcal{G}$.

(3) From almost commutativity 1.6.4 it follows, that γ_r is prime fixing. The rest is clear. $\qquad\square$

In particular ghosts are obtained in the following way.

COROLLARY 3.2.4. *Assume that in R there is a central element of degree one. Let $r \in R$ be normal. If for all units $u \in R_0^*$ the element ru is not central then for $\gamma = \gamma_r$ we have $\gamma^* \not\simeq 1_\mathcal{H}$, thus γ^* is a ghost.*

PROOF. By the assumption, $\gamma \notin \overline{\mathrm{Inn}}(R)$. $\square$

The proof of statement (2) in Proposition 3.2.3 shows that there is the following more general fact.

PROPOSITION 3.2.5. *Let γ be a graded algebra automorphism of R. Let x, y be points in $\mathbb{X}$ with corresponding homogeneous prime ideals P_x, P_y and simple objects S_x, S_y, respectively. If $\gamma(P_x) = P_y$ then $\gamma^*(S_x) \simeq S_y$. In particular, γ is prime fixing if and only if γ^* is point fixing.* $\square$

DEFINITION 3.2.6. An element γ of $\mathrm{Aut}(\mathbb{X})$ is called *liftable* to R, if there is a graded k-algebra automorphism ϕ of R such that ϕ_* represents γ.

Thus, a ghost γ is liftable to R if and only if its class in $\mathcal{G}$ lies in the image of the injective homomorphism $\mathrm{Aut}_0(R)/\overline{\mathrm{Inn}}(R) \longrightarrow \mathcal{G}$.

PROBLEM 3.2.7. Is any ghost liftable to R? Is $\mathrm{Aut}_0(R)$ modulo $\overline{\mathrm{Inn}}(R)$ generated by all the γ_r ($r \in R$ normal/prime)?

The main result of this chapter is the following.

THEOREM 3.2.8. *Let $R = \Pi(L, \sigma)$ with σ being efficient. Let $y \in \mathbb{X}$ with associated prime element $\pi_y \in R$ of degree d. Let $\gamma = \gamma_{\pi_y}$ be the associated algebra automorphism of R and γ^* the induced automorphism of $\mathcal{H}$. Then there is a natural isomorphism from the tubular shift σ_y to $\sigma^d \circ \gamma^*$, which is compatible with the natural transformations $1_\mathcal{H} \xrightarrow{y} \sigma_y$ and $1_\mathcal{H} \xrightarrow{\cdot \pi_y} \sigma^d \circ \gamma^*$.*

PROOF. Consider the proof of Proposition 3.2.3 (2) with $r = \pi_y = \pi_x$ (hence $u = 1$). Using the identity $\pi_y \cdot s = \gamma^{-1}(s) \cdot \pi_y$ for all $s \in R$, the lower exact sequence in the diagram (3.2.3) induces the commutative diagram

$$
\begin{array}{ccccccccc}
0 & \longrightarrow & \gamma^*(L) & \xrightarrow{\pi_y \cdot} & \gamma^*(L)(d) & \longrightarrow & \gamma^*(L_y) & \longrightarrow & 0 \\
  &   & \simeq \downarrow {\scriptstyle \gamma^{-1}} &   & \parallel &   & \simeq \downarrow &   &   \\
0 & \longrightarrow & L & \xrightarrow{\cdot \pi_y} & \gamma^*(L)(d) & \longrightarrow & L_y & \longrightarrow & 0,
\end{array}
$$

and hence $L \xrightarrow{\cdot \pi_y} \gamma^*(L)(d)$ yields an S_y-universal extension. The rest of the proof is completely analogue to the proof of 3.1.2. $\square$

COROLLARY 3.2.9. *If m is a positive integer such that π_y^m is central up to multiplication with some unit, then σ_y^m is naturally isomorphic to σ^{md}.*

PROOF. By the assumption, γ^m is an inner automorphism on R. $\square$

COROLLARY 3.2.10. *Let $x_1, \ldots, x_s \in \mathbb{X}$ and $\pi_{x_1}, \ldots, \pi_{x_s} \in R$ corresponding primes. Denote by $d_i = e(x_i)f(x_i)/\ell$ the degree of π_{x_i}. Let $m_1, \ldots, m_s$ be integers. Then*

$$(3.2.4) \qquad \sigma_{x_1}^{m_1} \circ \ldots \circ \sigma_{x_s}^{m_s} \simeq \sigma^{\sum_{i=1}^s m_i d_i} \circ (\gamma_{\pi_{x_1}}^*)^{m_1} \circ \ldots \circ (\gamma_{\pi_{x_s}}^*)^{m_s}.$$

PROOF. This follows from Theorem 3.2.8 together with the fact that the $\gamma_{\pi_{x_i}}^*$ commute with all $\gamma_{\pi_{x_j}}^*$ and all σ_{x_j} (by 0.4.8 and 1.6.2). $\square$

COROLLARY 3.2.11. *Let $\mathcal{G}'$ be the subgroup of the ghost group $\mathcal{G}$ generated by the automorphisms $\gamma_{\pi_x}^*$ ($x \in \mathbb{X}$). Then $\mathrm{Pic}(\mathbb{X}) \subset \langle \sigma \rangle \times \mathcal{G}'$.* □

COROLLARY 3.2.12. *Let $x_1, \ldots, x_s \in \mathbb{X}$ and $\pi_{x_1}, \ldots, \pi_{x_s} \in R$ corresponding primes. Denote by $d_i = e(x_i)f(x_i)/\ell$ the degree of π_{x_i}. Let $m_1, \ldots, m_s$ be integers. The following are equivalent:*

(1) $\sigma_{x_1}^{m_1} \circ \ldots \circ \sigma_{x_s}^{m_s} \simeq 1_{\mathcal{H}}$.

(2) $\sum_{i=1}^{s} m_i d_i = 0$ *and the graded automorphism $\gamma_{\pi_{x_1}}^{m_1} \circ \ldots \circ \gamma_{\pi_{x_s}}^{m_s}$ of R is in $\overline{\mathrm{Inn}}(R)$.* □

REMARK 3.2.13. Assume that there is a central element in $R = \Pi(L, \sigma)$ of degree one. This is equivalent to assuming that $\sigma = \sigma_x$ is an efficient tubular shift. Note that the existence of an efficient tubular shift σ_x is equivalent to the existence of a prime element of degree one in $\Pi(L, \sigma)$ (where σ is an arbitrary efficient automorphism). All this follows by Theorem 3.2.8 and Lemma 1.7.1. It follows that in the example in 1.1.13 (4) there is no prime element in R of degree one.

The existence of a central element of degree one implies the following:

(1) The subgroup $\overline{\mathrm{Inn}}(R)$ is generated by $\mathrm{Inn}(R)$ and automorphisms φ_a where a is the constant sequence with value $a \in Z(R)^*$.

(2) If $r \neq 0$ is normal, then $\gamma_r \in \overline{\mathrm{Inn}}(R)$ implies $\gamma_r \in \mathrm{Inn}(R)$.

COROLLARY 3.2.14. *Assume that there is a central element in R of degree one. Let $x_1, \ldots, x_s \in \mathbb{X}$ and $\pi_{x_1}, \ldots, \pi_{x_s} \in R$ corresponding primes. Denote by $d_i = e(x_i)f(x_i)/\ell$ the degree of π_{x_i}. Let $m_1, \ldots, m_s$ be integers. The following are equivalent:*

(1) $\sigma_{x_1}^{m_1} \circ \ldots \circ \sigma_{x_s}^{m_s} \simeq 1_{\mathcal{H}}$.

(2) $\sum_{i=1}^{s} m_i d_i = 0$ *and the graded automorphism $\gamma_{\pi_{x_1}}^{m_1} \circ \ldots \circ \gamma_{\pi_{x_s}}^{m_s}$ of R is inner.*

(3) *There is a unit $u \in R_0^*$ such that the element $u\pi_{x_1}^{m_1} \ldots \pi_{x_s}^{m_s}$ is a central element of degree zero in the graded quotient division ring $\mathrm{Quot}^{\mathbb{Z}}(R)$.*

(4) *There is a unit $u \in R_0^*$ such that the element $u\pi_{x_1}^{m_1} \ldots \pi_{x_s}^{m_s}$ lies in the centre of the function field $k(\mathbb{X})$.*

Note that we get as a special case a criterion for two tubular shifts σ_x and σ_y to be isomorphic.

PROOF. From the preceding corollary and remark we get the equivalence of (1) and (2), and also the equivalence of (2) and (3).

For the equivalence of (3) and (4) note that by the existence of the central unit of degree one we have $\mathrm{Quot}^{\mathbb{Z}}(R) = k(\mathbb{X})[T, T^{-1}]$, where T is a central variable of degree one. Then an element which lies in the centre of $k(\mathbb{X})$ also lies in the centre of $\mathrm{Quot}^{\mathbb{Z}}(R)$. □

We conclude the chapter by reformulating some results in the language of divisors.

3.2.15 (Divisors). (1) Denote by $\mathcal{N}_0^*$ the group of non-zero elements in $k(\mathbb{X})$ which are fractions of normal elements in R (of the same degree). Denote by $\mathrm{Div}(\mathbb{X})$ the abelian group of all formal sums of the form $\sum_{x \in \mathbb{X}} m_x x$ where $m_x \in \mathbb{Z}$, almost all zero. It follows from 1.6.3 that there is an exact sequence of groups

$$1 \longrightarrow R_0^* \longrightarrow \mathcal{N}_0^* \xrightarrow{\mathrm{div}} \mathrm{Div}(\mathbb{X}) \xrightarrow{\deg_\varepsilon} \mathbb{Z},$$

where $\deg_e(x) = e(x)f(x)/\ell$ and where div is induced by the correspondence $R\pi_x \mapsto x$ from 1.5.2. It follows from (3.2.4) that $\deg_e$ is surjective if and only if the positive integers $e(x)f(x)/\ell$ ($x \in \mathbb{X}$) generate the group $\mathbb{Z}$.

(2) Assume that there is a central element in R of degree one. Denote by $\mathcal{Z}_0^*$ the subgroup of $\mathcal{N}_0^*/R_0^*$ given by the classes which admit representatives lying in the centre of $k(\mathbb{X})$. The preceding discussion has shown that there is an exact sequence of abelian groups

$$0 \longrightarrow \mathcal{Z}_0^* \longrightarrow \mathcal{N}_0^*/R_0^* \longrightarrow \mathrm{Pic}_0(\mathbb{X}) \longrightarrow 0.$$

EXAMPLES 3.2.16. (1) The twisted polynomial case. Let $R = F[X;Y,\alpha]$, let x and y be the points corresponding to the primes X and Y, respectively. Since any prime ideal of height one different from RY is generated by a central element, for any points z_1, z_2 different from y we have $\sigma_{z_1} \simeq \sigma_{z_2}$ if and only if $d(z_1) = d(z_2)$. If α is not inner, then σ_x and σ_y are not isomorphic.

(2) The square roots case. Let $R = \mathbb{Q}\langle x, y, z \rangle$ be the graded factorial algebra from 1.7.12. It is easy to see that x (central), y and z are (up to multiplication with units) the only prime elements of degree one (using linear independence of x, y, z and Lemma 1.6.2). So there are precisely three unirational points, and the associated tubular shifts are pairwise non-isomorphic.

Further applications of the results of this section are discussed in Section 5.4 and in 5.7.2.

CHAPTER 4

Commutativity and multiplicity freeness

In this chapter we characterize the exceptional curves $\mathbb{X}$ which are commutative. We show that $\mathbb{X}$ is commutative if and only if it is multiplicity free. The proof of this is an application of the graded factoriality.

4.1. Finiteness over the centre

Before we characterize those exceptional curves which are commutative, we remark that exceptional curves in general are close to commutativity in the sense that they have a "large" centre. On the other hand, in the following sections it will be pointed out that they are commutative only in very special cases.

Let $\mathbb{X}$ be a homogeneous exceptional curve. It is well-known that the function field $k(\mathbb{X})$ is of finite dimension over its centre [7]. A similar result is true for the orbit algebras.

Let σ be efficient and $R = \Pi(L, \sigma)$.

PROPOSITION 4.1.1. *Let* $T = L \oplus \overline{L}$ *and* $S = \Pi(T, \sigma)$. *Then* S *is module-finite over its noetherian centre.*

PROOF. By [11] it is sufficient to show that S is a semiprime noetherian PI ring of global dimension two. This follows as in [7, Thm. 6.5]. □

COROLLARY 4.1.2. *Let* $R = \Pi(L, \sigma)$. *Then* R *is module-finite over its noetherian centre.*

PROOF. Let $e = \begin{pmatrix} 1_L & 0 \\ 0 & 0 \end{pmatrix}$, which is an element of degree zero in $S = \Pi(T, \sigma)$ with $e^2 = e$. Moreover, $R \simeq eSe \subset S$. Let C be the centre of S. Then eCe is commutative, noetherian and lies in the centre of $eSe = R$. Since S is a finitely generated C-module by the proposition, R is finitely generated over eCe. Then R is also finitely generated over its centre. □

COROLLARY 4.1.3. *Let* C *be the centre of* R. *The assignment* $P \mapsto P \cap C$ *is a bijection from the homogeneous prime ideals of height one in* R *onto the homogeneous prime ideals of height one in* C.

PROOF. We refer to graded versions of results in the literature: The surjectivity of the map follows by general properties of finite centralizing extensions [77, 10.2]. The injectivity follows from [36, 11.20] since in our situation cliques of prime ideals of height one are singletons. □

PROBLEM 4.1.4. *What is the geometric interpretation of this map? What is the structure and explicit form of the centre?*

See [18] for another approach studying this centre.

COROLLARY 4.1.5. *The centre of R is a graded normal domain of Krull dimension two.*

PROOF. It is sufficient to show that C is graded normal. Let rs^{-1} be a homogeneous element in the graded quotient field of C, which is a subfield of the graded quotient ring RX^{-1}, where X is the multiplicative set of normal elements. Then r and s are normal elements, and we cancel common prime factors in r and s and get a fraction $rs^{-1} = r's'^{-1} \in RX^{-1}$, where r' and s' have no common prime factor. If rs^{-1} is integral over C, then we see that s' has no prime factor, hence is a unit in R, and $rs^{-1} = r' \in R$. We get $r = r's$, and since r and s are central, also r' is central, that is, $r' \in C$: For each $x \in R$ we have

$$xr's = xr = rx = r'sx = r'xs,$$

and s can be canceled. $\square$

4.2. Commutativity of the coordinate algebra

If k is not algebraically closed it happens very rarely that the orbit algebra $\Pi(L, \sigma)$ is commutative[1]. It is shown in [**54**] (if the characteristic of k is different from two) that a small preprojective algebra is commutative if and only if there is some (commutative) finite field extension K/k such that the tame bimodule M is the Kronecker bimodule $_K(K \oplus K)_K$ or a $(4,1)$- or $(1,4)$-bimodule of a skew field of quaternions over K. So it is commutative only in very special cases. This result carries over to our type of orbit algebras. Another proof will be given by Theorem 4.3.5.

THEOREM 4.2.1. *Assume that the characteristic of k is different from two. Let $R = \Pi(L, \sigma)$, where σ is efficient. Then R is commutative if and only if there is a (commutative) finite field extension K/k such that as graded algebras either*

(a) *$R \simeq K[X,Y]$, the polynomial algebra graded by total degree; or*
(b) *$R \simeq K[X,Y,Z]/Q$, where $Q = Q(X,Y,Z)$ is an anisotropic quadratic form over K.*

Moreover, in case (a) M is the Kronecker bimodule $K \oplus K$ over K; in case (b) M is the bimodule $_K F_F$, where F is a skew field of quaternions over K. $\square$

Let Λ be the corresponding tame bimodule algebra. If we assume that k is the centre of Λ, then for the field extension K/k in theorem we have $K = k$.

Recall, that $\mathbb{X}$ is multiplicity free, if $e(x) = 1$ for all $x \in \mathbb{X}$. We call R *almost commutative*[2], if for all homogeneous r, $s \in R$ there is an $\alpha \in R_0^*$ such that $rs = \alpha sr$. Since in orbit case III the $(2,2)$-bimodule is simple, each rational point x (that is $f(x) = 1$), which always exists, has multiplicity $e(x) > 1$ (by 0.6.1). Hence, $\mathbb{X}$ is never multiplicity free in orbit case III.

PROPOSITION 4.2.2 (Almost commutativity). *R is almost commutative if and only if $\mathbb{X}$ is multiplicity free.*

PROOF. If $e(x) = 1$ for all $x \in \mathbb{X}$ then the prime elements are just the irreducible elements by 1.6.5 and 1.6.7. Then the almost commutativity follows by Proposition 1.6.3 since each non-zero homogeneous element is a product of irreducible elements.

---

[1]Commutativity is always meant in the usual and not in the graded sense.
[2]In the literature the term is also used with another meaning [**77**].

Conversely, if almost commutativity holds then each homogeneous prime ideal of height one (which is principal) is easily seen to be completely prime. Now apply Theorem 1.2.3 (3). $\qquad\square$

The preceding proposition will be strengthened in the following section.

4.3. Commutativity of the function field

As was pointed out by Ringel [**90**] there is a strange commutativity behaviour of the function field (see also [**23**]). For example ([**90**]) the bimodule ${}_\mathbb{R}\mathbb{H}_\mathbb{H}$ with noncommutative data leads to the commutative function field

$$\operatorname{Quot}\big(\mathbb{R}[U,V]/(U^2+V^2+1)\big)$$

whereas the bimodule ${}_\mathbb{Q}\mathbb{Q}(\sqrt{2},\sqrt{3})_{\mathbb{Q}(\sqrt{2},\sqrt{3})}$ with commutative data leads to the noncommutative function field given by the quotient division ring of

$$\mathbb{Q}\langle U,V\rangle/(UV+VU, V^2+2U^2-3).$$

(Compare also Proposition 1.7.12.) We will explain this effect in this section.

Recall that an exceptional curve $\mathbb{X}$ is called commutative if its function field $k(\mathbb{X})$ is commutative. The main result of this section is the following theorem. (Note that we allow also the weighted case in the theorem since it makes no difference.)

THEOREM 4.3.1. *Let $\mathbb{X}$ be an exceptional curve. The following statements are equivalent:*

(1) $\mathbb{X}$ *is commutative.*
(2) $\mathbb{X}$ *is multiplicity free.*
(3) *For each rational point $x \in \mathbb{X}$ we have $e(x) = 1$.*

In this case and if additionally char $k \neq 2$ there is some finite field extension K/k such that

- $k(\mathbb{X}) \simeq K(T)$ *if the numerical type of $\mathbb{X}$ is $\varepsilon = 1$; or*
- $k(\mathbb{X})$ *is isomorphic to the quotient field of $K[U,V]/(-aU^2 - bV^2 + ab)$ for some anisotropic quadratic form $-aX^2 - bY^2 + abZ^2$ over K if the numerical type of $\mathbb{X}$ is $\varepsilon = 2$.*

REMARK 4.3.2. (1) It follows (together with Theorem 5.3.4 and Proposition 5.5.1) that if char $k \neq 2$ and $\mathbb{X}$ is commutative then the ghost group $\mathcal{G}$ is trivial.

(2) In the theorem, K is the field $\operatorname{End}(L)$ and is the centre of $\mathcal{H}$.

Since the function field and the multiplicities are preserved by insertion of weights, it is sufficient to treat the homogeneous case.

Note that in case k is a finite field the equivalence (1)$\Leftrightarrow$(2) is given by Corollary 2.3.9. The implication (1)$\Rightarrow$(2) is given by 2.2.14. We give now another argument for this.

PROPOSITION 4.3.3. *Let $\mathbb{X}$ be homogeneous. Let $x \in \mathbb{X}$ be some point. Then as graded algebras, $\Pi(L,\sigma_x) \subset k(\mathbb{X})[T]$, where T is a central variable. In particular, if $k(\mathbb{X})$ is commutative, so is $\Pi(L,\sigma_x)$.*

PROOF. Let $\overline{\mathcal{H}} = \mathcal{H}/\mathcal{H}_0$. The natural transformation $1_\mathcal{H} \xrightarrow{x} \sigma_x$ induces a natural isomorphism $1_{\overline{\mathcal{H}}} \xrightarrow{\overline{x}} \overline{\sigma}_x$. Denote by $\overline{L}$ the class of L in $\overline{\mathcal{H}}$. Then there is an isomorphism of graded rings

$$\operatorname{End}_{\overline{\mathcal{H}}}(\overline{L})[T] \longrightarrow \bigoplus_{n \geq 0} \operatorname{Hom}_{\overline{\mathcal{H}}}(\overline{L}, \overline{\sigma}_x^n(\overline{L})), \quad fT^n \mapsto f * \overline{x}_{\overline{L}}^n$$

(where on the right hand side the multiplication and the power of $\overline{x}_{\overline{L}}$ are taken in the orbit algebra sense), and $\Pi(L, \sigma_x)$ embeds naturally.   □

COROLLARY 4.3.4. *Let* $\mathbb{X}$ *be homogeneous. Assume that the function field is commutative. Then* $\mathbb{X}$ *is multiplicity free. Moreover, there is some unirational point* x *such that the orbit algebra* $\Pi(L, \sigma_x)$ *is commutative graded factorial.*

PROOF. Take any point x and form the orbit algebra $R = \Pi(L, \sigma_x)$ with respect to the (not necessarily efficient) associated tubular shift σ_x. By the preceding proposition R is commutative. Moreover, by Serre's theorem 2.1.2, R is a homogeneous coordinate algebra for $\mathbb{X}$, and hence "classical" algebraic geometry shows that $\mathbb{X}$ is multiplicity free (see for example [55]). In particular, there exists some unirational point x and the assertion follows since σ_x is exhaustive.   □

The orbit algebras $R = \Pi(L, \sigma_x)$, where σ_x is exhaustive and where R is commutative, are described (in case char $k \neq 2$) in Theorem 4.2.1. From this we get the explicit form of the function fields as in Theorem 4.3.1. This explicit description follows again from the next theorem which provides also the proof for the implication $(3) \Rightarrow (1)$ in Theorem 4.3.1.

THEOREM 4.3.5. *Let* $\mathbb{X}$ *be homogeneous. Assume that for all rational points* $x \in \mathbb{X}$ *we have* $e(x) = 1$. *Then for each rational point* x *the orbit algebra* $\Pi(L, \sigma_x)$ *is commutative.*

Moreover, if $\varepsilon = 1$, *then there is a finite field extension* K/k *such that* $\Pi(L, \sigma_x) \simeq K[X, Y]$, *where* X *and* Y *are central variables of degree one. If* $\varepsilon = 2$ *and* char$(k) \neq 2$, *then there is a finite field extension* K/k *such that* $\Pi(L, \sigma_x) \simeq K[X, Y, Z]/(-aX^2 - bY^2 + abZ^2)$, *with* $a, b \in K^*$ *such that* $-aX^2 - bY^2 + abZ^2$ *is an anisotropic quadratic form over* K.

PROOF. There is a rational point x. By assumption x is unirational. Thus the associated tubular shift σ_x is efficient and the orbit algebra $R = \Pi(L, \sigma_x)$ graded factorial. We have $R = R_0\langle R_1 \rangle$ and $[R_1 : R_0] = \varepsilon + 1$. Moreover, $[R_n : R_0] = \varepsilon n + 1$. (Note that by Corollary 0.6.2 in case $\varepsilon = 1$ the underlying tame bimodule M is non-simple.)

FACT: *Let* u *be a non-zero element of* R, *homogeneous of degree one. Then* u *is prime.*

Namely, u is irreducible, hence by 1.3.3 a divisor of some prime element π, which because of $\deg(u) = 1$ is associated to a rational, hence multiplicity free point. It follows from 1.6.5 that u equals π up to some unit and hence is prime itself.

It follows that there are rational points y (and z) such that the prime elements π_x, π_y (and π_z) are linearly independent over R_0 and $R = R_0\langle \pi_x, \pi_y \rangle$ (in case $\varepsilon = 1$) or $R = R_0\langle \pi_x, \pi_y, \pi_z \rangle$ (in case $\varepsilon = 2$). Moreover, by 1.7.1 the prime π_x is central. We have to show

    (a) R_0 is commutative;
    (b) π_y (and π_z) commutes with each element from R_0;
    (c) $\pi_y \pi_z = \pi_z \pi_y$ (in case $\varepsilon = 2$).

We only discuss the case $\varepsilon = 2$ since the arguments for the case $\varepsilon = 1$ are similar and even easier. Since π_y and π_z are prime there is a unit $\alpha \in R_0^*$ such that $\pi_y \pi_z = \alpha \pi_z \pi_y$ (by 1.6.2). By the Fact above, $\pi_x + \pi_y$ is prime, therefore commutes with the prime π_z up to a unit, and $\alpha = 1$ follows. Let $f \in R_0$, $f \neq 0$. For $a \in R_0$, again by using the Fact above and considering the product $(\pi_x + a\pi_y)f$ it follows that $fa = a\gamma_y(f)$, where γ_y is the automorphism induced by the normal element π_y. In particular, for $a = 1$ we get $\gamma_y(f) = f$. It follows that R_0 is commutative and that π_y is central. Similarly, π_z is central. Thus, R is commutative.

With $K = R_0$ we get $R \simeq K[X, Y]$ in case $\varepsilon = 1$ and $R \simeq K[X, Y, Z]/Q$, where Q is a homogeneous quadratic polynomial, in case $\varepsilon = 2$. Assume $\mathrm{char}(k) \neq 2$. Then by factoriality Q is anisotropic over K and hence can be assumed to be of the form stated in the theorem. It follows also that the bimodule M is given as stated in Theorem 4.2.1. $\qquad\square$

From this, Theorem 4.3.1 follows immediately since $k(\mathbb{X})$ is the quotient division ring (of fractions of homogeneous elements of the same degree) of $\Pi(L, \sigma_x)$.

LEMMA 4.3.6. *Let $\mathbb{X}$ be homogeneous and $x \in \mathbb{X}$ a rational point. Then*

$$e(x) \leq \varepsilon \cdot [\mathrm{End}(L) : k].$$

PROOF. Since x is rational we have $[\mathrm{Ext}^1(S, L) : k] = \varepsilon \cdot [\mathrm{End}(L) : k]$, hence

$$e(x) = \frac{[\mathrm{Ext}^1(S, L) : k]}{[\mathrm{End}(S) : k]} = \frac{\varepsilon \cdot [\mathrm{End}(L) : k]}{[\mathrm{End}(S) : k]} \leq \varepsilon \cdot [\mathrm{End}(L) : k].$$

$\qquad\square$

Note that the curve may be defined over a field which is larger than k (for example, over the centre of the corresponding bimodule algebra) and that the formula also holds with this larger field instead of k.

EXAMPLE 4.3.7. Let M be the $\mathbb{Q}$-$\mathbb{Q}(\sqrt{2}, \sqrt{3})$-bimodule $\mathbb{Q}(\sqrt{2}, \sqrt{3})$. Let x be a unirational point and $R = \Pi(L, \sigma_x)$. By 1.7.12

$$R = \mathbb{Q}\langle X, Y, Z \rangle / (XY - YX, XZ - ZX, YZ + ZY, Z^2 + 2Y^2 - 3X^2).$$

Since this algebra is not commutative it follows from Theorem 4.3.5 that there is a rational point of multiplicity greater than one. We determine such a point explicitly. Denote the images of X, Y and Z by x, y and z, respectively. Then x is central, y and z are normal (but not central). The element $u = x - y$ is irreducible but not normal, hence not prime. But u is divisor of a prime $\pi = \pi_p$. Then p is a rational point with $e(p) > 1$.

More precisely, (up to multiplication with a unit in $\mathbb{Q}^*$) we have $\pi = x^2 - y^2$: In fact, by the preceding lemma we have $e(p) = 2$, hence $\deg(\pi) = 2$. By Proposition 1.3.3 there is $v \in R$ (of degree 1) such that $vu = \pi$. Moreover, $x^2 - y^2 = (x - y) \cdot (x + y)$ lies in the centre of R. (Note that it follows easily from $z(x - y) = (x + y)z$ that $x - y$ and $x + y$ induce isomorphic simple cokernels.) Then $v(x^2 - y^2) = vu(x + y) = \pi(x + y)$. Since π is prime, it follows that π divides $x^2 - y^2$, and since both have the same degree they are associated.

COROLLARY 4.3.8. *Let $\mathbb{X}$ be homogeneous and $R = \Pi(L, \sigma)$ with σ efficient. The following are equivalent:*

(1) *$k(\mathbb{X})$ is commutative.*
(2) *R is commutative.*

(2') *R is almost commutative.*

(3) *For each homogeneous prime ideal P in R of height one the localization R_P^0 is a local (equivalently, a semiperfect) ring.*

PROOF. Recall Proposition 4.2.2 and Proposition 2.2.15. □

PROBLEM 4.3.9. Let $\mathbb{X}$ be a homogeneous exceptional curve. Is the category $\mathcal{H} = \mathrm{coh}(\mathbb{X})$ uniquely determined (up to equivalence) by the function field $k(\mathbb{X})$? This is only clear for commutative function fields.

In the context of this section Problem 2.3.10 becomes interesting again:

PROBLEM 4.3.10. Find a formula for the skewness $s(\mathbb{X})$ in terms of the multiplicity function e. From such a formula Theorem 4.3.1 should be derived as a special case.

REMARK 4.3.11. In case $s(\mathbb{X}) = 2$ the existence of a rational point x such that $e(x) = s(\mathbb{X})$ follows directly from Theorem 4.3.1.

REMARK 4.3.12. The function field and the multiplicities are also related by a fundamental exact sequence. For this sequence one has to consider the Grothendieck category $\mathrm{Qcoh}(\mathbb{X}) = \frac{\mathrm{Mod}^{\mathbb{Z}}(R)}{\mathrm{Mod}_0^{\mathbb{Z}}(R)}$, the quotient category modulo the Serre subcategory formed by the $\mathbb{Z}$-graded torsion modules.

The injective hull Q of the line bundle L is a generic sheaf (corresponding to the generic Λ-module, where Λ is the associated bimodule algebra). Moreover, Q is the injective hull of each line bundle and the endomorphism ring $\mathrm{End}(Q)$ is the function field $k(\mathbb{X})$. (Compare [56, Lemma 14].)

For each $x \in \mathbb{X}$ denote by S_x^{ω} the Prüfer sheaf, which is the direct limit of all $S_x^{(n)}$ (the indecomposable sheaf of length n concentrated in x).

There is the short exact sequence in $\mathrm{Qcoh}(\mathbb{X})$

$$0 \longrightarrow L \longrightarrow Q \longrightarrow \bigoplus_{x \in \mathbb{X}} \bigoplus_{e(x)} S_x^{\omega} \longrightarrow 0$$

involving the multiplicities $e(x)$. This sequence already appeared in [90, Prop. 5.2].

CHAPTER 5

Automorphism groups

Knowledge of the automorphism group of a curve contributes enormously to a better understanding of the geometry. This chapter is devoted to the determination of automorphism groups. By the mechanism of insertion of weights this problem will be reduced essentially to the homogeneous case (see Section 6.3), which we treat now.

The group $\text{Aut}(\mathcal{H})$ consists essentially of $\text{Pic}(\mathbb{X})$ and $\text{Aut}(\mathbb{X})$, where the latter group is given by the geometric automorphisms and the ghosts. Over an algebraically closed field the Picard group is just $\mathbb{Z}$, but this does not hold in general, as will be shown in Theorem 5.4.1. So far, the best theorem on the structure of $\text{Pic}(\mathbb{X})$ we have is Theorem 3.2.8, which also gives a hint as to what ghosts look like.

In Proposition 5.1.4 we prove that $\text{Aut}(\mathbb{X})$ is isomorphic to the factor group of the automorphism group of the underlying bimodule M modulo the inner bimodule automorphisms, which is useful for explicit calculations. Sections 5.3 and 5.4 constitute the main part of this chapter, where we analyse the case of a non-simple bimodule M. In this case it is easy to desribe the graded factorial coordinate algebra explicitly, which will be helpful in the determination of $\text{Aut}(\mathbb{X})$. This graded algebra is of the form

$$F[X; Y, \alpha, \delta],$$

where F is a finite-dimensional skew field, X is a central variable, $\alpha \in \text{Aut}(F/k)$ and $\delta : F \longrightarrow F$ is a $(\alpha, 1)$-derivation such that for all $f \in F$ the following relation holds

$$Yf = \delta(f)X + \alpha(f)Y.$$

We concentrate on the case $\delta = 0$, thus $R = F[X; Y, \alpha]$ is a graded twisted polynomial algebra. If r is the order of α modulo inner automorphisms, then α induces a ghost automorphism α_* of order r. In the special case where α induces a generator of the Galois group $\text{Gal}(Z/k)$ (with Z being the centre of F) we determine $\text{Aut}(\mathbb{X})$ explicitly. Our knowledge of the prime spectrum of R can be used to distinguish ghosts from geometric automorphisms.

As an application we get a formula for the Auslander-Reiten translation. One might expect that τ is given by degree shift by -2, that is, by σ_x^{-2}. While this is true on objects, it can fail on morphisms; the correct formula is

$$\tau = \sigma_x^{-1} \circ \sigma_y^{-1} = \sigma_x^{-2} \circ \alpha_*^{-1}.$$

We will treat also the quaternion case. Over the real numbers this is the case when $M = {}_{\mathbb{R}}\mathbb{H}_{\mathbb{H}}$, where $\mathbb{H}$ is the skew field of quaternions over $\mathbb{R}$. Together with our results on the non-simple bimodules we describe the homogeneous exceptional curves explicitly when $k = \mathbb{R}$, the field of real numbers. It turns out that such a curve over $\mathbb{R}$ can be identified with a quotient of the Riemann sphere modulo an involution, possibly equipped with additional structure, and that the geometric automorphism group is just the group of directly conformal homeomorphisms on the Riemann sphere compatible with the involution and respecting the additional structure.

We conclude the chapter by exhibiting an interesting homogeneous exceptional curve over $k = \mathbb{Q}$ whose automorphism group is isomorphic to the Klein four group.

5.1. The automorphism group of a homogeneous curve

Now we treat the automorphism group in the homogeneous case. We will see, that calculations of the automorphism group will lead us into (noncommutative) Galois theory.

5.1.1 (Bimodule automorphisms). Let $M = {}_F M_G$ be a bimodule over the skew fields F and G, k acting centrally, with all data finite dimensional over k. We always assume $M \neq 0$. Define the group $\mathrm{Aut}(M) = \mathrm{Aut}_k({}_F M_G)$ to be the set of all triples $(\varphi_F, \varphi_M, \varphi_G)$, where $\varphi_F \in \mathrm{Aut}(F/k)$, $\varphi_G \in \mathrm{Aut}(G/k)$, $\varphi_M : M \longrightarrow M$ is k-linear and bijective, and for all $f \in F$, $g \in G$ and $m \in M$ we have

$$\varphi_M(fmg) = \varphi_F(f)\varphi_M(m)\varphi_G(g).$$

Composition and inverse are built componentwise, the neutral element is given by $(1_F, 1_M, 1_G)$. Note that projection onto the middle component, $(\varphi_F, \varphi_M, \varphi_G) \mapsto \varphi_M$ is injective. There is an alternative description: Consider the k-category consisting of two objects with endomorphism ring F and G, respectively, and with non-zero Hom-space only in one direction, which is given by M. Then an automorphism of the bimodule M is just an autoequivalence of this category.

5.1.2 (Inner automorphisms). An element $(\varphi_F, \varphi_M, \varphi_G) \in \mathrm{Aut}(M)$ is called *inner*, if there are $f \in F^*$, $g \in G^*$ such that for all $x \in F$, $y \in G$, $m \in M$ we have $\varphi_F(x) = f^{-1}xf$, $\varphi_G(y) = g^{-1}yg$ and $\varphi_M(m) = f^{-1}mg$. The subgroup of all inner automorphisms is denoted by $\mathrm{Inn}(M) = \mathrm{Inn}_k({}_F M_G)$, the factor group by $\mathrm{Out}(M) = \mathrm{Out}_k({}_F M_G) = \mathrm{Aut}(M)/\mathrm{Inn}(M)$.

5.1.3. Each element $(\varphi_F, \varphi_M, \varphi_G) \in \mathrm{Aut}(M)$ defines a k-algebra automorphism on the hereditary algebra $\Lambda := \begin{pmatrix} G & 0 \\ M & F \end{pmatrix}$ in the obvious way, and conversely; then, the triple is inner if and only if the induced k-algebra automorphism is inner.

PROPOSITION 5.1.4. *Let $\mathbb{X}$ be a homogeneous exceptional curve with underlying tame bimodule $M = {}_F M_G$. Then*

$$\mathrm{Aut}(\mathbb{X}) \simeq \mathrm{Out}(M).$$

PROOF. Denote by $\overline{L}$ the indecomposable bundle such that there is an irreducible map from L to $\overline{L}$ as in 1.1.2. Then $M = \mathrm{Hom}(L, \overline{L})$. Let φ be an autoequivalence of $\mathcal{H} = \mathrm{coh}\,\mathbb{X}$ fixing the structure sheaf L. Then φ also fixes $\overline{L}$. Therefore, by restriction φ induces an autoequivalence of the full subcategory $\{L, \overline{L}\}$, hence an element of $\mathrm{Aut}(M)$. Moreover, with a theorem of Eilenberg and Watts [8, II.2.3], the functor φ is isomorphic to the identity on $\mathcal{H}$ if and only if its restriction is isomorphic to the identity on $\{L, \overline{L}\}$, that is, if and only if the induced automorphism on the bimodule is inner.

Conversely, any element ϕ in $\mathrm{Aut}(M)$ induces an automorphism ϕ of the bimodule algebra Λ. This induces an autoequivalence ϕ_* of $\mathrm{mod}(\Lambda)$ by sending a right Λ-module N to $N_{\phi^{-1}}$ (observe the -1 here, ensuring that the assignment $\phi \mapsto \phi_*$ is covariant). This induces an autoequivalence of the triangulated category $D^b(\Lambda)$, and since $D^b(\Lambda) = D^b(\mathbb{X})$ this finally induces an autoequivalence (which we also denote by ϕ_*) of $\mathcal{H}$. Now ϕ_* fixes L since it is easy to see that the corresponding indecomposable summand of Λ is preserved under the automorphism $\phi: \Lambda \longrightarrow \Lambda$. Moreover, ϕ is inner if and only if ϕ_* on mod(Λ) is isomorphic to the identity (compare [**8**, II.5]). These constructions are mutually inverse. $\square$

REMARK 5.1.5. Let $M = {}_F M_G$ be a tame bimodule over k. Denote by K its centre (compare 0.5.5). Clearly, $\mathrm{Aut}_K(M) \subset \mathrm{Aut}_k(M)$ and $\mathrm{Inn}_K(M) = \mathrm{Inn}_k(M)$. Moreover, there is an exact sequence

$$1 \longrightarrow \mathrm{Aut}_K(M) \longrightarrow \mathrm{Aut}_k(M) \xrightarrow{\ \rho\ } \mathrm{Gal}(K/k),$$

where ρ is defined by restricting φ_F and φ_G to K. It follows that the factor group $\mathrm{Aut}_k(\mathbb{X})/\mathrm{Aut}_K(\mathbb{X})$ can be embedded into $\mathrm{Gal}(K/k)$.

5.2. The structure of Aut($\mathcal{H}$)

5.2.1 (Orbit cases IIIa and IIIb). Let $\mathbb{X}$ be a homogeneous exceptional curve with hereditary category $\mathcal{H}$. We would like to analyse the structure of $\mathrm{Aut}(\mathcal{H})$. In order to do this we have to refine the definition of orbit case III. By definition of this case there is no $\sigma \in \mathrm{Aut}_0(\mathcal{H})$ with $\sigma(L) = \overline{L}$ (see 1.1.5). But is it possible that there is a $\sigma \in \mathrm{Aut}(\mathcal{H})$ with this property? If there is such an automorphism, we call this case orbit case IIIb, if not, we call it case IIIa. In other words, in case IIIa each efficient automorphism is transitive. Of course, if in case III the endomorphism skew fields of line bundles in two different Auslander-Reiten orbits are non-isomorphic, then we are in case IIIa.

PROBLEM 5.2.2. Is orbit case IIIb non-empty, that is, does there exist a tame bimodule over some field belonging to this case?

Recall that $\mathcal{O}$ is the $\mathrm{Aut}(\mathcal{H})$-orbit and $\mathcal{O}_0$ is the $\mathrm{Aut}_0(\mathcal{H})$-orbit of L. Note that only in orbit case IIIb there is a difference between $\mathcal{O}$ and $\mathcal{O}_0$. Obviously, in any orbit case there exists a transitive automorphism: in case IIIb by definition, in all other cases there is even a point fixing one.

Fix a transitive $\sigma \in \mathrm{Aut}(\mathcal{H})$ (which is unique up to an automorphism of $\mathbb{X}$). Let $\phi \in \mathrm{Aut}(\mathcal{H})$. Then there is a (unique) $n \in \mathbb{Z}$ such that $\sigma^n(\phi(L)) \simeq L$, that is, $\sigma^n \circ \phi \in \mathrm{Aut}(\mathbb{X})$. In other words, each $\phi \in \mathrm{Aut}(\mathcal{H})$ is a composition of a power of σ and some element of $\mathrm{Aut}(\mathbb{X})$. Therefore, $\mathrm{Aut}(\mathcal{H})$ consists of the following types of automorphisms:

- the geometric automorphisms of the curve
- the tubular shifts
- the ghosts
- one transitive automorphism (if not already in the Picard group)

As usual, denote by $\mathcal{G}$ the ghost group of $\mathbb{X}$. Recall that $\mathrm{Pic}_0(\mathbb{X}) = \mathrm{Pic}(\mathbb{X}) \cap \mathrm{Aut}(\mathbb{X})$, which is $\mathrm{Pic}(\mathbb{X}) \cap \mathcal{G}$.

PROPOSITION 5.2.3. *Let $\mathbb{X}$ be a homogeneous exceptional curve. Assume, that there is an exhaustive automorphism σ lying in $\mathrm{Pic}(\mathbb{X})$, and assume that we are not in orbit case IIIb. Then there are split exact sequences of groups*

$$1 \longrightarrow \mathrm{Pic}(\mathbb{X})/\mathrm{Pic}_0(\mathbb{X}) \longrightarrow \mathrm{Aut}(\mathcal{H})/\mathcal{G} \longrightarrow \mathrm{Aut}(\mathbb{X})/\mathcal{G} \longrightarrow 1,$$

where $\mathrm{Pic}(\mathbb{X})/\mathrm{Pic}_0(\mathbb{X}) \simeq \mathbb{Z}$ *with generator induced by* σ, *and*

$$1 \longrightarrow \langle \sigma \rangle \longrightarrow \mathrm{Aut}_0(\mathcal{H}) \longrightarrow \mathcal{G} \longrightarrow 1.$$

Moreover, $\mathrm{Aut}(\mathcal{H})/\mathrm{Aut}_0(\mathcal{H}) \simeq \mathrm{Aut}(\mathbb{X})/\mathcal{G}$.

PROOF. Since we exclude orbit case IIIb, σ is transitive. Denote the class of an automorphism ϕ modulo $\mathcal{G}$ by $[\phi]$. It is easy to see that $[\phi] \mapsto [\sigma^n \circ \phi]$, where $\sigma^n(\phi(L)) \simeq L$, is a well-defined homomorphism $\mathrm{Aut}(\mathcal{H})/\mathcal{G} \longrightarrow \mathrm{Aut}(\mathbb{X})/\mathcal{G}$ admitting a section and having kernel $\mathrm{Pic}(\mathbb{X})/\mathrm{Pic}_0(\mathbb{X})$. For the second sequence note, that for $\phi \in \mathrm{Aut}_0(\mathcal{H})$ and any point x we have $\phi \circ \sigma_x \circ \phi^{-1} = \sigma_x$, hence $\phi \circ \sigma \circ \phi^{-1} = \sigma$, thus the map $\phi \mapsto \sigma^n \circ \phi$ such that $\sigma^n(\phi(L)) \simeq L$ is a homomorphism. The final assertion is clear by considering the map $\mathrm{Aut}(\mathbb{X})/\mathcal{G} \longrightarrow \mathrm{Aut}(\mathcal{H})/\mathrm{Aut}_0(\mathcal{H})$, which is induced by the inclusion. □

Proposition 5.2.3 will be extended to weighted curves in Proposition 6.3.4.

5.3. The twisted polynomial case

We now discuss the case, where the underlying bimodule is non-simple, moreover of the form $M(F, \alpha)$. That is, we assume that the derivation δ is trivial. In the special case where the automorphism $\alpha \in \mathrm{Aut}(F/k)$ is a generator of the Galois group of the centre Z/k we determine the automorphism group $\mathrm{Aut}(\mathbb{X})$ completely. It follows in particular that α induces a generator of the ghost group. In the next section we will deduce a formula for the Auslander-Reiten translation.

5.3.1. Let $M = M(F, \alpha)$. Recall that $\Pi(L, \sigma) \simeq F[X; Y, \alpha]$, where X is central. Let x and y be the points corresponding to the prime ideals generated by X and Y, respectively, and let σ_x and σ_y be the associated tubular shifts.

Modulo inner automorphisms α has finite order r. There is some $u \in \mathrm{Fix}(\alpha)^*$ such that $\alpha^r(f) = u^{-1}fu$ for all $f \in F$. Recall that for an element $f \in F$ the norm of f is given by $N(f) = \alpha^{r-1}(f) \cdots \alpha(f)f$.

Denote by $\mathrm{Gal}(F/k)$ the factor group $\mathrm{Aut}(F/k)/\mathrm{Inn}(F/k)$. By the Skolem-Noether theorem [**30**], restriction to the centre $Z = Z(F)$ induces a monomorphism $\mathrm{Gal}(F/k) \subset \mathrm{Gal}(Z/k)$. Note that if k is the field of real numbers or a finite field (hence F commutative), then $\mathrm{Gal}(F/k)$ is cyclic.

5.3.2. The automorphism $\alpha \in \mathrm{Aut}(F/k)$ induces an automorphism of the bimodule M, given by $(f, g) \mapsto (\alpha(f), \alpha(g))$, which we denote also by α. Denote by $\alpha_* \in \mathrm{Aut}(\mathbb{X})$ the induced automorphism of the curve, as described in the proof of 5.1.4.

Let γ be the graded algebra automorphism of $R = F[X; Y, \alpha]$ given by $rY = Y\gamma(r)$ for all $r \in R$. Then α_* coincides with $(\gamma^{-1})_* = \gamma^*$ (as defined in 3.2.1). In fact, it is easy to see that these automorphisms coincide on the full subcategory $\{L, \overline{L} = L(1)\}$, and by the argument given in the proof of Proposition 5.1.4 they coincide also on $\mathcal{H}$.

From the description of the prime elements in 1.7.7 the next lemma follows immediately:

LEMMA 5.3.3. α *induces a prime fixing automorphism of* R. *Accordingly* α_* *is a ghost of order* r. □

For the automorphism group we treat a special, cyclic Galois case.

THEOREM 5.3.4. *Let Z be the centre of F and $\alpha \in \mathrm{Aut}(F/k)$ such that its image in $\mathrm{Gal}(F/k)$ generates the group $\mathrm{Gal}(Z/k)$ of order r.*

(1) *Let $r = 1$. Then $\mathrm{Aut}(\mathbb{X})$ is canonically isomorphic to $\mathrm{PGL}_2(Z)$.*

(2) *Let $r \geq 2$. Then $\mathrm{Aut}(\mathbb{X})$ is generated by*

- *the automorphism α_* induced by α, which generates the ghost group $\mathcal{G}$ and is of order r;*
- *transformations of the form $Y \mapsto aY$ for $a \in Z^*$; they are liftable to R; two transformations $Y \mapsto aY$ and $Y \mapsto bY$ (with $a, b \in Z^*$) give the same automorphism on $\mathbb{X}$ if and only if $N(a) = N(b)$;*
- *if $r = 2$ additionally by the (non-liftable[1]) automorphism exchanging X for uY (and Y for $u^{-1}X$).*

PROOF. We assume $r \geq 2$, since the case $r = 1$ is easy. Let M be the underlying bimodule as described above. By 5.1.4, $\mathrm{Aut}(\mathbb{X})$ is given by $\mathrm{Out}(M) = \mathrm{Aut}(M)/\mathrm{Inn}(M)$. Elements of $\mathrm{Aut}(M)$ are given by triples $(\varphi_1, \varphi_M, \varphi_2)$ with φ_1, $\varphi_2 \in \mathrm{Aut}(F/k)$ and $\varphi_M \in \mathrm{End}_k(M)$ bijective such that $\varphi_M(f_1 m f_2) = \varphi_1(f_1)\varphi_M(m)\varphi_2(f_2)$ for all f_1, $f_2 \in F$ and $m \in M$. Note that modulo inner bimodule automorphisms we can assume that φ_1 and φ_2 are powers of α. Recall that there is the automorphism induced by α, given as (α, α, α), where on ${}_F M = F \oplus F$ it is given by $\alpha(x, y) = (\alpha(x), \alpha(y))$. By abuse of notation, we denote this automorphism of M by the same letter α. We have $\alpha_*^r = 1$. Note that in case $\varphi_1 = 1$, φ_M is given by an invertible matrix,

$$\varphi_M(x, y) = (x, y) \cdot \begin{pmatrix} a & b \\ c & d \end{pmatrix}$$

with $a, b, c, d \in F$ such that $ad - bc \neq 0$. Calculating $\varphi_M((1, 0) \cdot f)$ and $\varphi_M((0, 1) \cdot f)$ for all $f \in F$, one gets in case $\varphi_2 = 1$ that the matrix is diagonal with entries lying in Z^* (exploit the existence of $f \in Z$ such that $\alpha(f) \neq f$). Similarly, an automorphism of the form $(1, \varphi_M, \alpha^j)$ with $1 \leq j \leq r - 1$ is only possible for $r = 2$ and $j = 1$, which leads to the matrices of the form $\begin{pmatrix} 0 & a \\ b & 0 \end{pmatrix}$ with $a, b \in Z^*$ (using $f \in Z$ such that $\alpha^j(f) \neq f$). Moreover, a diagonal matrix is inner if and only if it is of the form $\begin{pmatrix} ab & 0 \\ 0 & a\alpha(b) \end{pmatrix}$ with a, b in Z^*. The norm N induces a map, assigning a diagonal matrix $\begin{pmatrix} a & 0 \\ 0 & b \end{pmatrix}$ the value $N(a^{-1}b)$, and, involving Hilbert's Theorem 90 (applied to the cyclic Galois extension Z/K where $K = Z \cap \mathrm{Fix}(\alpha)$, see [63]), such a matrix is mapped to the identity if and only if it is of the above diagonal form with twist. Moreover, up to inner automorphisms we can assume $a = 1$, hence we have a transformation $Y \mapsto bY$. Since $(bY)f = \alpha(f)(bY)$ for all $f \in F$, this transformation extends to a graded k-algebra automorphism of $R = F[X; Y, \alpha]$, mapping $X^i Y^j$ to $N_j(b) X^i Y^j$. Moreover, for $N(b) \neq 1$ this induces a geometric element in $\mathrm{Aut}(\mathbb{X})$: otherwise it would be prime fixing, in particular it would fix the prime element $X^r + uY^r$, and then $N(b) = N_r(b) = 1$ would follow. □

REMARK 5.3.5. (1) With the assumptions and notations of the theorem let $r \geq 2$ and $K = Z \cap \mathrm{Fix}(\alpha)$. Let U be the subgroup of Z^* of elements a with

---

[1]This automorphism is induced by the canonical isomorphisms of graded algebras $\Pi(L, \sigma_x) \simeq F[X; Y, \alpha] \simeq F[uY; X, \alpha^{-1}] \simeq \Pi(L, \sigma_y)$.

$N(a) = 1$. Then

$$\text{Aut}(\mathbb{X}) \simeq \begin{cases} (Z^*/U \rtimes \mathbb{Z}_2) \rtimes \langle \alpha_* \rangle & r = 2; \\ Z^*/U \rtimes \langle \alpha_* \rangle & r > 2; \end{cases}$$

where $\mathcal{G} = \langle \alpha_* \rangle$ is the cyclic ghost group of order r. Let $K_+ = N(Z^*) \subset K^*$. Then N induces an isomorphism of groups $Z^*/U \simeq K_+$. But note that the action of K_+ on $\mathbb{X}$ is not given explicitly.

(2) In case $r = 2$, even though X is central in R and Y is not, the localizations $R^0_{(X)}$ and $R^0_{(Y)}$ are isomorphic (by mapping XY^{-1} to uYX^{-1}). (Compare 2.2.5.)

EXAMPLES 5.3.6. (1) The theorem can be applied for $k = \mathbb{R}$ to the bimodules $M = \mathbb{R} \oplus \mathbb{R}$ and $M = \mathbb{H} \oplus \mathbb{H}$ with $\alpha = 1$ and $r = 1$. Also, all $(2,2)$-bimodules over finite fields are captured by the theorem.

(2) Let $k = \mathbb{R}$ and $M = \mathbb{C} \oplus \overline{\mathbb{C}}$, where $\mathbb{C}$ acts on the right hand side on the second component via complex conjugation α. Then $r = 2$, and the elements of $\text{Aut}(\mathbb{X})$ are given by $Y \mapsto rY$ $(r > 0)$, by "inversion" $X \leftrightarrow Y$ and by the ghost α_*.

(3) Let $k = \mathbb{Q}(\mathbf{i})$ and $F = k(\sqrt[4]{2})$, let α be the automorphism $\sqrt[4]{2} \mapsto \mathbf{i}\sqrt[4]{2}$ and let $M = M(F, \alpha)$. Then $\text{Aut}(\mathbb{X})$ consists of the ghost α_* of order 4 and the automorphisms $Y \mapsto aY$ $(a \in F^*)$. Here an element $a \in F^*$ cannot always be represented (modulo the group U of elements of norm 1) by an element in $\text{Fix}(\alpha) = \mathbb{Q}(\mathbf{i})$. For example, $N(\sqrt[4]{2}) = -2$ cannot coincide with the norm of an element in $\mathbb{Q}(\mathbf{i})^*$.

5.4. On the Auslander-Reiten translation as functor

We continue to study the non-simple bimodule case where we still assume that for the derivation we have $\delta = 0$. Hence the orbit algebra is of the form $F[X; Y, \alpha]$. As before, let x and y be the points corresponding to the prime elements X and Y, respectively.

THEOREM 5.4.1. Let $R = \Pi(L, \sigma_x) = F[X; Y, \alpha]$ with $\alpha \in \text{Aut}(F/k)$. Let r be the order of α modulo inner automorphisms. Then
(1) As elements of $\text{Aut}\,\mathcal{H}$, $\sigma_x^{-1} \circ \sigma_y = \alpha_*$.
(2) $\text{Pic}(\mathbb{X}) = \langle \sigma_x, \sigma_y \rangle \simeq \mathbb{Z} \times \mathbb{Z}_r$ and $\text{Pic}_0(\mathbb{X}) = \langle \alpha_* \rangle \simeq \mathbb{Z}_r$.
(3) The Auslander-Reiten translation τ acts on elements of $\text{End}(L)$ like $\sigma_x^{-1} \circ \sigma_y^{-1}$.

PROOF. (1) Let γ be the automorphism of the graded algebra R given by $r\pi_y = \pi_y \gamma(r)$ (which coincides on R_0 with α^{-1}). Then by definition, $\gamma^* = \alpha_*$, and the formula follows from 3.2.8.

(2) As before, let $u \in \text{Fix}(\alpha)^*$ such that $\alpha^r(f) = u^{-1}fu$ for all $f \in F$. Then, $\pi_x \in R$ is central, π_y is normal such that $u\pi_y^r$ is central, and every other prime element in R is central. Moreover by (1), for $1 \le j \le r-1$, the automorphisms σ_y^j differ from the powers of σ_x. Hence $\text{Pic}(\mathbb{X}) = \langle \sigma_x, \sigma_y \rangle \simeq \mathbb{Z} \times \mathbb{Z}_r$ follows from Corollary 3.2.9. Moreover, $\text{Pic}_0(\mathbb{X})$ is generated by $\sigma_x^{-1} \circ \sigma_y = \alpha_*$.

(3) Let $\mathfrak{m}$ be the homogeneous Jacobson radical of the graded local ring R. From the diagram of the Koszul complex (see 2.1.8)

$$
\begin{array}{ccccccccc}
0 & \longrightarrow & R(-2) & \xrightarrow{(Y\cdot\ (-X)\cdot)^t} & R(-1)\oplus R(-1) & \xrightarrow{(X\cdot\ Y\cdot)} & R & \longrightarrow & R/\mathfrak{m} \\
& & \downarrow{\scriptstyle g} & & \downarrow{\scriptstyle\begin{pmatrix} a & b \\ c & d \end{pmatrix}} & & \downarrow{\scriptstyle f} & & \downarrow \\
0 & \longrightarrow & R(-2) & \xrightarrow{(Y\cdot\ (-X)\cdot)^t} & R(-1)\oplus R(-1) & \xrightarrow{(X\cdot\ Y\cdot)} & R & \longrightarrow & R/\mathfrak{m}
\end{array}
$$

it follows that $b = 0 = c$, $a = f(-1)$, $d = \alpha^{-1}(f)(-1)$, and $g = \alpha^{-1}(f)(-2)$. By sheafification we get for any $f \in \mathrm{End}(L)$ a diagram of almost split sequences

$$
\begin{array}{ccccccccc}
\mu: & 0 & \longrightarrow & \tau L = L(-2) & \longrightarrow & L(-1)\oplus L(-1) & \longrightarrow & L & \longrightarrow 0 \\
& & & \downarrow{\scriptstyle \sigma_x^{-2}\alpha^{-1}(f)} & & \downarrow & & \downarrow{\scriptstyle f} & \\
\mu: & 0 & \longrightarrow & \tau L = L(-2) & \longrightarrow & L(-1)\oplus L(-1) & \longrightarrow & L & \longrightarrow 0.
\end{array}
$$

It follows that (on classes) $\sigma_x^{-2}\alpha^{-1}(f)\cdot\mu = \mu\cdot f$, for any $f \in \mathrm{End}(L)$. On the other hand, as can be derived from [**75**, Lemma 3], $\tau(f)\cdot\mu = \mu\cdot f$, and $\tau(f) = \sigma_x^{-2}\alpha^{-1}(f)$ follows. Now apply (1). $\qquad\square$

Let S_x and S_y be the simple objects associated to π_x and π_y, respectively. Mapping $f \in \mathrm{End}(L)$ to its fibre map f_x induces an isomorphism $\mathrm{End}(S_x) \simeq \mathrm{End}(L) = F$, and similarly for S_y. Then the following is easy to see.

COROLLARY 5.4.2. *On elements of* $\mathrm{End}(S_x)$ *and* $\mathrm{End}(S_y)$ *the Auslander-Reiten translation* τ *acts like* α *and* $\alpha^{-1} \in \mathrm{Aut}(F/k)$, *respectively.* $\qquad\square$

COROLLARY 5.4.3. *On the tube* $\mathcal{U}_x$ $(\mathcal{U}_y)$ *the tubular shift* σ_x (σ_y) *coincides with* α^* $(\alpha_*,$ *respectively) and hence does not coincide, in case* $r \geq 2$, *with the identity functor on this tube.* $\qquad\square$

COROLLARY 5.4.4. *Assume, that the powers of* α_* *are the only ghosts (which is true, for example, under the assumptions of Theorem 5.3.4). Then as elements in* $\mathrm{Aut}(\mathcal{H})$,

$$\tau = \sigma_x^{-1}\circ\sigma_y^{-1} = \sigma_x^{-2}\circ\alpha^*.$$

In particular, $\tau \in \mathrm{Pic}(\mathbb{X})$.

PROOF. On objects τ acts like σ_x^{-2}, thus $\sigma_x^2\circ\tau$ is a ghost, which must be α_*^{-1} by the theorem and the assumption. $\qquad\square$

Because of the identity $\tau = \sigma_x^{-1}\circ\sigma_y^{-1}$ the following is immediate:

COROLLARY 5.4.5. τ *is the identity functor on the length categories* $\mathcal{U}_z$ *for every point* $z \neq x, y$. $\qquad\square$

PROBLEM 5.4.6. Is $\tau \in \mathrm{Pic}(\mathbb{X})$ true for any exceptional curve? Does the equation $\tau = \sigma_x^{-1}\circ\sigma_y^{-1}$ hold under the weaker assumptions of the preceding theorem? Find a general functorial formula for τ.

PROBLEM 5.4.7. Extend the results of this and the preceding section to arbitrary non-simple bimodules, that is, to arbitrary $\alpha \in \mathrm{Aut}(F/k)$ and to the case where $\delta \neq 0$. What can be said in case of a simple bimodule?

5.5. The quaternion case

Let k be a field of characteristic different from two. Let a, $b \in k^*$ and let $F = \left(\frac{a,b}{k}\right)$ be an algebra of quaternions over k, that is, a k-algebra on generators $\mathbf{i}$ and $\mathbf{j}$ subject to the relations

$$\mathbf{ji} = -\mathbf{ij}, \ \mathbf{i}^2 = a, \ \mathbf{j}^2 = b.$$

Moreover, we assume that F is a skew field. Let M be the bimodule ${}_k F_F$. Let $F_0 = k\mathbf{i} \oplus k\mathbf{j} \oplus k\mathbf{ji}$ be the quadratic space of pure quaternions, where the quadratic form is given by the restricted norm form $q = -aX^2 - bY^2 + abZ^2$. Let $\mathrm{SO}(F_0)$ be the group of all isometries of this quadratic space with determinant 1 (see [**61**]).

PROPOSITION 5.5.1. *Let $\mathbb{X}$ be a homogeneous exceptional curve, where the underlying bimodule is given as above. We have an isomorphism $\mathrm{Aut}(\mathbb{X}) \simeq \mathrm{SO}(F_0)$ which induces the canonical action of $\mathrm{SO}(F_0)$ on the projective spectrum of the coordinate algebra $R = k[X, Y, Z]/(-aX^2 - bY^2 + abZ^2)$. Each automorphism of $\mathbb{X}$ is geometric.*

PROOF. We exhibit the proof which is given in [**58**] for $k = \mathbb{R}$. We have to calculate the (outer) automorphisms of the bimodule M. For each $a \in F^*$ denote by $\iota_a : F \longrightarrow F$ the inner automorphism given by $\iota_a(f) = a^{-1}fa$ for all $f \in F$. Each $\varphi \in \mathrm{Aut}(M)$ has the form $\varphi = (1, \varphi, \iota_a)$, where $\varphi(f) = \varphi(1)a^{-1}fa$. We obtain a surjection $F^* \rtimes F^* \longrightarrow \mathrm{Aut}(M)$ with kernel $1 \rtimes k^*$, hence $\mathrm{Aut}(M) \simeq F^* \rtimes F^*/k^*$. Since every inner automorphism of the bimodule M is of the form $x \mapsto \alpha^{-1}xa$ for some $\alpha \in k^*$, $a \in F^*$, there is a surjection $k^* \rtimes F^* \longrightarrow \mathrm{Inn}(M)$ inducing an isomorphism $\mathrm{Inn}(F) \simeq k^* \rtimes F^*/k^*$. Hence $\mathrm{Out}(M) \simeq F^*/k^* \simeq \mathrm{SO}(F_0)$ (see [**61**]). By the correspondence between the basis $\mathbf{i}$, $\mathbf{j}$, $\mathbf{k} = \mathbf{ji}$ of F_0 and the elements x, y, z in R (as described in [**54**, 4.3]), we see how an element of $\mathrm{SO}(F_0)$ acts on (prime) elements of degree one in R, and this action extends uniquely to an automorphism of the graded k-algebra R.   □

5.6. The homogeneous curves over the real numbers

In this section we apply and illustrate our results on the automorphism group in the special situations where $k = \mathbb{R}$ is the field of real numbers. There are (up to duality) only five tame bimodules, which are listed in the following table. The corresponding graded factorial coordinate algebras and automorphism groups $\mathrm{Aut}(\mathbb{X})$ were determined in the preceding sections. In the table, γ denotes complex conjugation, I the inversion $z \mapsto 1/z$ (explained below). Note that in the "classical" case $\mathbb{C} \oplus \mathbb{C}$ complex conjugation occurs, since we consider it as bimodule over $\mathbb{R}$. Moreover, in this case complex conjugation induces a geometric automorphism.

	M	$R = \Pi(L, \sigma_x)$	$\mathrm{Aut}(\mathbb{X})$
1.	${}_{\mathbb{R}}\mathbb{H}_{\mathbb{H}}$	$\mathbb{R}[X, Y, Z]/(X^2 + Y^2 + Z^2)$	$\mathrm{SO}_3(\mathbb{R})$
2.	${}_{\mathbb{R}}\mathbb{R}_{\mathbb{R}} \oplus {}_{\mathbb{R}}\mathbb{R}_{\mathbb{R}}$	$\mathbb{R}[X, Y]$	$\mathrm{PGL}_2(\mathbb{R})$
3.	${}_{\mathbb{C}}\mathbb{C}_{\mathbb{C}} \oplus {}_{\mathbb{C}}\mathbb{C}_{\mathbb{C}}$	$\mathbb{C}[X, Y]$	$\mathrm{PGL}_2(\mathbb{C}) \rtimes \langle \gamma_* \rangle$
4.	${}_{\mathbb{H}}\mathbb{H}_{\mathbb{H}} \oplus {}_{\mathbb{H}}\mathbb{H}_{\mathbb{H}}$	$\mathbb{H}[X, Y]$, X, Y central	$\mathrm{PGL}_2(\mathbb{R})$
5.	${}_{\mathbb{C}}\mathbb{C}_{\mathbb{C}} \oplus {}_{\mathbb{C}}\mathbb{C}_{\overline{\mathbb{C}}}$	$\mathbb{C}[X, \overline{Y}]$, X central, $Yz = \overline{z}Y$	$(\mathbb{R}_+ \rtimes \langle I_* \rangle) \rtimes \langle \gamma_* \rangle$

TABLE 5.1. The real homogeneous curves

5.6.1. Let $R = \Pi(L, \sigma)$, where σ is an efficient automorphism. The points of $\mathbb{X}$ are the prime elements in R. In each of the five cases we list the prime elements (up to units), the endomorphism skew fields of the corresponding simple objects S_x and the symbol data $\begin{pmatrix} d(x) \\ f(x) \end{pmatrix}$. We call a point $x \in \mathbb{X}$ real (complex, quaternion) if the endomorphism ring of S_x is $\mathbb{R}$ ($\mathbb{C}$, $\mathbb{H}$, respectively). We call the property of x being real, complex or quaternion, respectively, also the *colouring* of x.

(1) $\mathbb{R}[X, Y, Z]/(X^2 + Y^2 + Z^2) = \mathbb{R}[x, y, z]$.
- $ax + by + cz$, $(a, b, c) \neq (0, 0, 0)$; $\mathbb{C}$; $\begin{pmatrix} 1 \\ 1 \end{pmatrix}$

Hence $\mathbb{X}$ can be identified with $\mathbb{S}^2/\pm 1$, the 2-sphere modulo antipodal points. This is homeomorphic to $\mathbb{P}^1(\mathbb{C})/\mathbb{Z}_2$, the Riemann sphere modulo the fixed-point free involution (given by $z \mapsto -1/\bar{z}$ on $\mathbb{P}^1(\mathbb{C})$). There are no real points.

(2) $\mathbb{R}[X, Y]$.
- X, $Y + \alpha X$ $\alpha \in \mathbb{R}$; $\mathbb{R}$; $\begin{pmatrix} 1 \\ 1 \end{pmatrix}$.
- $(Y + zX)(Y + \bar{z}X)$ $z \in \mathbb{C} \setminus \mathbb{R}$; $\mathbb{C}$; $\begin{pmatrix} 2 \\ 2 \end{pmatrix}$.

Hence $\mathbb{X} = \mathbb{P}^1(\mathbb{C})/\mathbb{Z}_2$ (identifying X, $Y + \alpha X$, $(Y + zX)(Y + \bar{z}X)$ with the class of ∞, α, z, respectively) where here $\mathbb{Z}_2$ is generated by the involution (given by $z \mapsto \bar{z}$) having fixed points (= real points). We have two regions, the boundary (= real points) having symbol data $\begin{pmatrix} 1 \\ 1 \end{pmatrix}$ and the inner points are complex having symbol data $\begin{pmatrix} 2 \\ 2 \end{pmatrix}$.

(3) $\mathbb{C}[X, Y]$.
- X, $Y + zX$ $z \in \mathbb{C}$; $\mathbb{C}$; $\begin{pmatrix} 1 \\ 1 \end{pmatrix}$.

Here, $\mathbb{X} = \mathbb{P}^1(\mathbb{C})$, the Riemann sphere.

(4) $\mathbb{H}[X, Y]$.
- X, $Y + \alpha X$ $\alpha \in \mathbb{R}$; $\mathbb{H}$; $\begin{pmatrix} 1 \\ 1 \end{pmatrix}$.
- $(Y + zX)(Y + \bar{z}X)$ $z \in \mathbb{C} \setminus \mathbb{R}$; $\mathbb{C}$; $\begin{pmatrix} 2 \\ 1 \end{pmatrix}$.

Here $\mathbb{X} = \mathbb{P}^1(\mathbb{C})/\mathbb{Z}_2$ (as in case (2)), but the boundary is coloured quaternion.

(5) $\mathbb{C}[X, \overline{Y}]$.
- X, Y; $\mathbb{C}$; $\begin{pmatrix} 1 \\ 1 \end{pmatrix}$
- $Y^2 - \alpha X^2 = (Y - \sqrt{\alpha}X)(Y + \sqrt{\alpha}X)$ $0 < \alpha \in \mathbb{R}$; $\mathbb{R}$; $\begin{pmatrix} 2 \\ 1 \end{pmatrix}$
- $Y^2 - \alpha X^2$ $0 > \alpha \in \mathbb{R}$; $\mathbb{H}$; $\begin{pmatrix} 2 \\ 2 \end{pmatrix}$
- $(Y^2 - zX^2)(Y^2 - \bar{z}X^2)$ $z \in \mathbb{C} \setminus \mathbb{R}$; $\mathbb{C}$; $\begin{pmatrix} 4 \\ 2 \end{pmatrix}$.

In this case, the points of $\mathbb{X}$ are in ono-to-one correspondence with the elements of $\mathbb{P}^1(\mathbb{C})/\mathbb{Z}_2$ (mapping X, Y, $Y^2 - \alpha X^2$ ($0 \neq \alpha \in \mathbb{R}$), $(Y^2 -$

$zX^2)(Y^2 - \bar{z}X^2)$ $(z \in \mathbb{C} \setminus \mathbb{R})$ to the class of ∞, 0, α, z in $\mathbb{P}^1(\mathbb{C})/\mathbb{Z}_2$, respectively). The boundary is coloured in a more complicated fashion as in the preceding cases and is indicated in Figure 5.1.

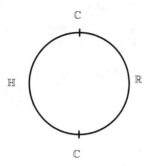

FIGURE 5.1

Thus, in each of the five cases $\mathbb{X}$ can be identified with the Riemann sphere $\mathbb{P}^1(\mathbb{C})$ or a quotient of it modulo some involution plus some additional structure. Non-boundary points are always complex.

PROPOSITION 5.6.2 ([**58**]). *Let $\mathbb{X}$ be a homogeneous exceptional curve over $k = \mathbb{R}$. The geometric automorphism group of $\mathbb{X}$ is canonically isomorphic to the group of directly conformal homeomorphisms on the Riemann sphere respecting the involution and the colouring.*

PROOF. The geometric elements of $\mathrm{Aut}(\mathbb{X})$ for all real cases are explicitly described in 5.3.4 and 5.5.1 as certain invertible 2×2-matrices. Thus these elements act in a natural way as Möbius transformations on the Riemann sphere, which induces a surjective homomorphism from $\mathrm{Aut}(\mathbb{X})$ to the group of conformal homeomorphisms respecting the involution and the colouring [**46**] (ghosts mapped to the identity), which establishes the isomorphism as claimed. Moreover, having identified $\mathbb{X}$ with the Riemann sphere as above, this action coincides with the action of $\mathrm{Aut}(\mathbb{X})$ on $\mathbb{X}$, with the only exception that in case $\mathbb{C}[X, \overline{Y}]$ a transformation of the form $X \mapsto aX$ corresponds to the Möbius transformation $z \mapsto a^2 z$. $\square$

In case $k = \mathbb{R}$ Proposition 5.2.3 reads as follows.

PROPOSITION 5.6.3. *Let $\mathbb{X}$ be a homogeneous exceptional curve over $k = \mathbb{R}$. Let $\mathcal{G}$ be the ghost group. Then the group $\mathrm{Pic}(\mathbb{X})/\mathcal{G}$ acts simply transitive on the set of all (isomorphism classes of) line bundles and there is a split exact sequence of groups*

$$1 \longrightarrow \mathrm{Pic}(\mathbb{X})/\mathcal{G} \longrightarrow \mathrm{Aut}(\mathcal{H})/\mathcal{G} \longrightarrow \mathrm{Aut}(\mathbb{X})/\mathcal{G} \longrightarrow 1.$$ $\square$

The weighted version is given in 6.3.5. We restate Corollary 5.4.4:

PROPOSITION 5.6.4. *Let $k = \mathbb{R}$. Only in case $M = \mathbb{C} \oplus \overline{\mathbb{C}}$ the ghost group $\mathcal{G}$ is non-trivial. In this case $\mathcal{G}$ is generated by the automorphism γ of order two induced by complex conjugation. Let x and y be the unirational points. Then*

$$\tau = \sigma_x^{-1} \circ \sigma_y^{-1} = \sigma_x^{-2} \circ \gamma.$$ $\square$

5.7. Homogeneous curves with finite automorphism group

Over a finite field the automorphism group of a homogeneous curve $\mathbb{X}$ is finite (by 5.1.4). But also in characteristic zero the automorphism group of a homogeneous curve may be finite.

EXAMPLE 5.7.1. Let F/k be commutative with $[F : k] = 4$ and $M = {}_kF_F$. Then it is easy to see that $\mathrm{Out}(M) \simeq \mathrm{Gal}(F/k)$.

EXAMPLE 5.7.2. Let M be the $\mathbb{Q}$-$\mathbb{Q}(\sqrt{2}, \sqrt{3})$-bimodule $\mathbb{Q}(\sqrt{2}, \sqrt{3})$. With the notations as in 4.3.7, the elements y and z in $R = \Pi(L, \sigma_x)$ are not central up to a unit, hence define (non-trivial) ghosts of order 2 (by 3.2.4). In fact they generate $\mathrm{Aut}(\mathbb{X})$ which is isomorphic to the Klein four group $\mathbb{V}_4$. Each automorphism of $\mathbb{X}$ (except the identity) is a ghost.

In the rest of this section we will elaborate a more complicated example which we will meet again later when discussing a tubular curve of index three, see 8.3.1.

EXAMPLE 5.7.3. Let $k = \mathbb{Q}$ and $F = \left(\frac{-1, -1}{\mathbb{Q}}\right)$ be the skew field of quaternions over $\mathbb{Q}$ on generators $\mathbf{i}$, $\mathbf{j}$ with relations $\mathbf{i}^2 = -1 = \mathbf{j}^2$, $\mathbf{ij} = -\mathbf{ji}$, $K = \mathbb{Q}(\sqrt{-3}, \sqrt{2})$ and M be the bimodule ${}_K(K \oplus K)_F$ with the canonical K-action, and where the F-action on M is defined by

$$(x, y) \cdot \mathbf{i} = \frac{1}{\sqrt{-3}}(\sqrt{2}x + y, x - \sqrt{2}y), \ (x, y) \cdot \mathbf{j} = (y, -x)$$

for all $x, y \in K$. Since $K \not\simeq F$, the bimodule M is simple. Let $\mathbb{X}$ be the homogeneous exceptional curve over this bimodule.

PROPOSITION 5.7.4. $\mathrm{Aut}(\mathbb{X}) \simeq \mathbb{V}_4$, the Klein four group, and every automorphism is geometric. Moreover, there is a rational point x, which is fixed by all automorphisms.

PROOF. We have to calculate the (outer) automorphisms $(\gamma, \varphi, \delta)$ of the bimodule M. Since every $\mathbb{Q}$-automorphism of F is inner, modulo inner bimodule automorphisms we can assume $\delta = 1_F$. Moreover, for γ we have the possibilities, that γ is the identity, or $\gamma = \alpha$, β or $\beta\alpha$, where $\mathrm{Gal}(K/\mathbb{Q}) = \langle \alpha, \beta \rangle$, with

$$\begin{array}{ccc} \alpha(\sqrt{2}) & = & \sqrt{2} \\ \alpha(\sqrt{-3}) & = & -\sqrt{-3} \end{array} \quad \text{and} \quad \begin{array}{ccc} \beta(\sqrt{2}) & = & -\sqrt{2} \\ \beta(\sqrt{-3}) & = & \sqrt{-3} \end{array}.$$

For $\gamma = 1_K$, using that $\varphi((1, 0)\mathbf{i}) = \varphi(1, 0)\mathbf{i}$ and $\varphi((1, 0)\mathbf{j}) = \varphi(1, 0)\mathbf{j}$ one immediately gets, that φ is represented by a scalar matrix with non-zero entries in K, and hence induces the identity modulo inner automorphisms. It is easy to see, that modulo inner, the only automorphism with $\gamma = \alpha$ is $(\alpha, \widetilde{\alpha}, 1)$, where

$$\widetilde{\alpha}(x, y) = (\alpha(x), \alpha(y)) \cdot \mathbf{j},$$

which is not inner since $\alpha \neq 1$. Namely, if $(\alpha, \varphi, 1)$ is an additional automorphism, then $(\alpha, \widetilde{\alpha}, 1) \circ (\alpha, \varphi, 1) = (1, \widetilde{\alpha}\varphi, 1)$. Moreover, $(\beta, \widetilde{\beta}, 1)$ is an automorphism, where

$$\widetilde{\beta}(x, y) = \frac{\sqrt{2}}{\sqrt{-3}} \cdot (\beta(x), \beta(y)) \cdot \mathbf{j} + \frac{1}{\sqrt{-3}} \cdot (\beta(x), \beta(y)),$$

and this is modulo inner the only automorphism with $\gamma = \beta$. Composing these two automorphisms gives modulo inner the unique automorphism with $\gamma = \beta\alpha$, and we get the Klein four group.

We have to show that each automorphism is geometric: For example, we have $\widetilde{\alpha}(\sqrt{2},\sqrt{-3}) = -(\sqrt{-3},\sqrt{2})$, and a simple calculation shows, that there are no $a \in K^*$, $f \in F^*$ such that $-(\sqrt{-3},\sqrt{2}) = a(\sqrt{2},\sqrt{-3})f$. Hence $\widetilde{\alpha}$ moves the point associated to $(\sqrt{2},\sqrt{-3})$.

Let $m = (1,1)$. Then $\widetilde{\alpha}(m) = m\mathbf{j}$ and $\widetilde{\beta}(m) = m\mathbf{i}$, hence $\widetilde{\alpha}$ and $\widetilde{\beta}$ fix the point associated to m (compare Lemma 0.6.3). $\qquad\square$

PROPOSITION 5.7.5. *Let* $\mathbb{X}$ *be as above. Each rational point has multiplicity 2 or 4, and there are rational points in both cases. In particular, there exists an efficient tubular shift* σ_x.

PROOF. By Lemma 0.6.1 the multiplicity of any rational point is 2 or 4. Let $m = (\sqrt{2}+\sqrt{-3}+\sqrt{-6},1) \in M$. Then the intersection $Km \cap mF$ is of dimension two over $\mathbb{Q}$. (In fact, one calculates that the elements $(a - \frac{1}{2}b\sqrt{-6}) \cdot m$ (with a, $b \in \mathbb{Q}$) lie in the intersection.) By Lemma 0.6.1 the corresponding point x has multiplicity 2, and then the associated shift σ_x is efficient. Similarly, for $m = (1,1)$ the intersection $Km \cap mF$ is of dimension one over $\mathbb{Q}$, hence the multiplicity of the corresponding point is 4. $\qquad\square$

REMARK 5.7.6. Let $\mathbb{X}$ be as before, and let x be a rational point. The preceding proposition shows in particular that $\mathrm{End}(S_x)$ is commutative, and hence $e^*(x) = 1$. Since there are infinitely many rational points (by [89]) we conclude from 2.3.5 that $s(\mathbb{X}) = 4$, and that there are only finitely many rational points x with $e(x) = 2$.

Part 2

The weighted case

CHAPTER 6

Insertion of weights

In this chapter we show how results for homogeneous curves can be extended to weighted curves. We concentrate here on the graded factoriality and on the automorphism groups. As a technical tool we will make use of the p-cycle construction by Lenzing.

6.1. p-cycles

In this section we briefly describe the p-cycle construction by Lenzing [**68**]. It is related to the concept of a parabolic structure [**103**]. It follows from the construction that each exceptional curve arises from a *homogeneous* exceptional curve by insertion of weights into a finite number of points. This is, together with its inverse process, the perpendicular calculus [**35**], the most important tool in order to reduce problems to the homogeneous (=unweighted) case. We will consider the following problems:

- construction of graded factorial coordinate algebras for $\mathcal{H}$ in the weighted cases by describing an equivalent process of insertion of weights into prime elements of the graded factorial algebras in the underlying homogeneous cases (see 6.2.4).
- determination of the automorphism group $\mathrm{Aut}(\mathbb{X})$ and the Picard group $\mathrm{Pic}(\mathbb{X})$ in the weighted cases (see 6.3.1 and 6.3.3).

6.1.1 (*p-cycles in x*). Let $\mathbb{X}$ be an exceptional curve with associated hereditary category $\mathcal{H}$. Let $x \in \mathbb{X}$ be a homogeneous point and $p \geq 2$ be an integer ("weight"). We use the natural transformation $1_{\mathcal{H}} \xrightarrow{x} \sigma_x$. A p-cycle (concentrated) in x is an integer indexed sequence

$$ E = (E_i, x_i)_{i \in \mathbb{Z}} = [\cdots \to E_0 \xrightarrow{x_0} E_1 \xrightarrow{x_1} E_2 \to \cdots \to E_{p-1} \xrightarrow{x_{p-1}} E_0(x) \to \cdots] $$

of morphisms $E_i \xrightarrow{x_i} E_{i+1}$ in $\mathcal{H}$ such that $E_{i+p} = E_i(x)$ and $x_{i+p} = \sigma_x(x_i)$ for all $i \in \mathbb{Z}$ and each composition $x_{i+p-1} \circ x_{i+p-2} \circ \cdots \circ x_i$ coincides with the natural morphism x_{E_i} for all $i \in \mathbb{Z}$. We write

$$ E = [E_0 \xrightarrow{x_0} E_1 \xrightarrow{x_1} E_2 \to \cdots \to E_{p-1} \xrightarrow{x_{p-1}} E_0(x)]. $$

A morphisms f between p-cycles $E = (E_i, x_i)$ and $F = (F_i, y_i)$ in x consists of morphisms $f_i : E_i \longrightarrow F_i$ such that $f_{i+1} \circ x_i = y_i \circ f_i$ and $f_{i+p} = f_i(x)$ for all $i \in \mathbb{Z}$.

The category of all p-cycles in x is denoted by $\overline{\mathcal{H}} = \mathcal{H}\begin{pmatrix} p \\ x \end{pmatrix}$ and is easily seen to be abelian and noetherian.

6.1.2 (Inclusion). Let $\overline{\mathcal{H}} = \mathcal{H} \begin{pmatrix} p \\ x \end{pmatrix}$. There is the full exact embedding $j : \mathcal{H} \longrightarrow \overline{\mathcal{H}}$ given by
$$j(E) = [\, E = E = E = \cdots = E \xrightarrow{x_E} E(x) \,]$$
We have a left adjoint $\ell : \overline{\mathcal{H}} \longrightarrow \mathcal{H}$ and a right adjoint $r : \overline{\mathcal{H}} \longrightarrow \mathcal{H}$ of j given by $\ell(E_i, x_i) = E_{p-1}$ and $r(E_i, x_i) = E_0$.

6.1.3 (Simple objects). The simple objects in $\overline{\mathcal{H}}$ are given (via the inclusion j) by the simple objects in $\mathcal{H}$ concentrated in points y different from x, together with the following p-cycles in x:

$$S_1 = [\, 0 \to 0 \to \cdots \to 0 \to S \to 0 \,]$$
$$S_2 = [\, 0 \to 0 \to \cdots \to S \to 0 \to 0 \,]$$
$$\cdots$$
$$S_{p-1} = [\, 0 \to S \to \cdots \to 0 \to 0 \to 0 \,]$$
$$S_p = [\, S \to 0 \to \cdots \to 0 \to 0 \to S(x) \,]$$

where $S \in \mathcal{H}$ is simple and concentrated in x. The simple objects $S_1, \ldots, S_p$ are exceptional with $\mathrm{End}(S_i) = \mathrm{End}(S)$ and $\mathrm{Ext}^1(S_{i+1}, S_i) \neq 0$.

Let $L \in \mathcal{H}$ be a special line bundle. Then $j(L) \in \overline{\mathcal{H}}$ is also special, since $\mathrm{Hom}(j(L), S_i) \neq 0$ if and only if $i = p$.

6.1.4 (Reduction of weights/perpendicular calculus). Let $\mathcal{S}$ be the subset $\{S_2, \ldots, S_p\}$ of the simple objects concentrated in x except S_1, and denote by $\langle \mathcal{S} \rangle$ the extension closure of $\mathcal{S}$ in $\overline{\mathcal{H}}$. This forms a localizing Serre subcategory in $\overline{\mathcal{H}}$, and the functor $\ell : \overline{\mathcal{H}} \longrightarrow \mathcal{H}$ induces an equivalence between the quotient category $\overline{\mathcal{H}}/\langle \mathcal{S} \rangle$ and $\mathcal{H}$.

The right perpendicular category $\mathcal{S}^\perp$ formed in $\overline{\mathcal{H}}$ is equivalent to $\mathcal{H}$.

6.1.5 (Tubular shifts). Let $\overline{\mathcal{H}}$ be the category of p-cycles in x.

(1) On $\overline{\mathcal{H}}$ the tubular shift $\overline{\sigma}_x$ associated to x corresponds to the natural transformation $1_{\overline{\mathcal{H}}} \longrightarrow \overline{\sigma}_x$, indicated by the following diagram

$$
\begin{array}{ccccccccc}
\overline{E} = [& E_0 \xrightarrow{x_0} E_1 & \longrightarrow \cdots \longrightarrow & E_{p-2} \xrightarrow{x_{p-2}} & E_{p-1} \xrightarrow{x_{p-1}} & E_0(x) &] \\
& \downarrow \quad x_0 \downarrow \quad x_1 \downarrow & & x_{p-2} \downarrow & x_{p-1} \downarrow & x_0(x) \downarrow & \\
\overline{\sigma}_x \overline{E} = [& E_1 \xrightarrow{x_1} E_2 & \longrightarrow \cdots \longrightarrow & E_{p-1} \xrightarrow{x_{p-1}} & E_0(x) \xrightarrow{x_0(x)} & E_1(x) &]
\end{array}
$$

(2) For every point $y \in \mathbb{X}$ the associated tubular shift $\sigma_y : \mathcal{H} \longrightarrow \mathcal{H}$ extends in an obvious way componentwise to an automorphism on $\overline{\mathcal{H}}$, again denoted by σ_y, and which is for $y \neq x$ just the tubular shift on $\overline{\mathcal{H}}$ associated to $y \in \overline{\mathbb{X}}$. It is clear that the relations in $\mathrm{Aut}(\overline{\mathcal{H}})$ between the σ_y ($y \in \mathbb{X}$) are the same as in $\mathrm{Aut}(\mathcal{H})$. Moreover, $(\overline{\sigma}_x)^p = \sigma_x$.

(3) More generally, each automorphism on $\mathcal{H}$ which is point fixing can be extended componentwise onto $\overline{\mathcal{H}}$ (compare 0.4.8). In particular this holds for an efficient automorphism σ defined on the sheaf category over the underlying homogeneous curve.

6.1.6 (Reduction to the homogeneous case). We say that $\mathcal{H} \begin{pmatrix} p_1, \ldots, p_t \\ x_1, \ldots, x_t \end{pmatrix}$ (which is defined inductively for pairwise different points $x_1, \ldots, x_t$) arises by insertion of a finite number of weights from $\mathcal{H}$. Let $\mathcal{H}$ be some abelian k-category. The following are equivalent [**68**, Thm. 5.1]:

(1) $\mathcal{H}$ is equivalent to the hereditary category associated to an exceptional curve $\mathbb{X}$.

(2) $\mathcal{H}$ arises by insertion of a finite number of weights from the hereditary category associated to a homogeneous exceptional curve.

6.2. Insertion of weights into central primes

We discuss the notion of insertion of weights on the level of the coordinate algebra. We will extend Serre's theorem to weighted situations, at least in special cases. Generally it is not difficult to construct projective coordinate algebras in the weighted cases, even if there exists no tilting object. In the case of an exceptional curve, with the theorem by Artin and Zhang [**2**, Thm. 4.5] and by using a tilting bundle T, forming an orbit algebra with respect to T, the construction of a projective coordinate algebra is easy.

PROPOSITION 6.2.1. *Let $\mathbb{X}$ be an arbitrary exceptional curve, and let $T \in \mathcal{H}_+$ be a tilting bundle. Let σ_x be the tubular shift at some point $x \in \mathbb{X}$. Then the pair (T, σ_x) is ample. Hence $\Pi(T, \sigma_x)$ is a projective coordinate algebra for $\mathbb{X}$.*

PROOF. The tilting bundle T defines the torsion class $\mathcal{T} = \{X \in \mathcal{H} \mid \mathrm{Ext}^1(T, X) = 0\}$. By Serre duality, $\mathcal{H}_0 \subset \mathcal{T}$. Thus, if $F \in \mathcal{H}$, then by 0.4.6 and Serre duality we have $F(n) \in \mathcal{T}$ for sufficiently large n. Then $F(n)$ is a quotient of T^s for some $s > 0$ (see [**40**]). This implies the first property of ampleness, the second follows again with 0.4.6 and Serre duality. $\square$

A different and more general construction of an ample pair is described in [**88**, IV.4] (and [**87**, Appendix A]), which leads in our setting to an ample pair (L, σ), where L is a (special) line bundle and σ is a composition of certain tubular shifts.

The projective coordinate algebras constructed in either of these two ways are not practical for our considerations. We would like to preserve the graded factoriality, starting from the homogeneous case.

6.2.2 (Insertion of weights into central primes). Let $\mathbb{X}$ be a homogeneous exceptional curve with sheaf category $\mathcal{H}$ and $R = \Pi(L, \sigma)$ with σ efficient, which is (positively) graded by the abelian group H ($= \mathbb{Z}$). Let $P = R\pi$ be a homogeneous prime ideal of height one, where we assume that π is central and of degree d. (More generally, R may be a k-algebra graded by an abelian group H, and π a central, homogeneous element.) Let $p > 1$ be a "weight". Define $\overline{R} = R[\pi^{1/p}] = R[T]/(T^p - \pi)$, where T is a central variable. Denote $\overline{H} = H[\frac{d}{p}]$ and by $\overline{\pi} \in \overline{R}$ the image of T. Then $\overline{R}$ is an $\overline{H}$-graded algebra with $\deg(\overline{\pi}) = \overline{h} = \frac{d}{p}$. If moreover $(H, \leq)$ is an ordered abelian group (defined by the positive cone $H_+ = \{h \in H \mid h \geq 0\}$) such that R is positively H-graded ($R_h \neq 0$ only for $h \geq 0$), then $\overline{R}$ is positively $\overline{H}$-graded, where the ordering on $\overline{H}$ is defined by the positive cone $\overline{H}_+ = \{h + n\overline{h} \mid h \in H_+, \; n \geq 0\}$. This can be iterated.

EXAMPLE 6.2.3. Let $\mathbf{p} = (p_1, \ldots, p_t)$ and $\mathbf{d} = (d_1, \ldots, d_t)$ be sequences of positive integers. Then the group $\mathbb{Z}\left[\frac{d_1}{p_1}, \ldots, \frac{d_t}{p_t}\right]$ is denoted by $\mathbb{L}(\mathbf{p}, \mathbf{d})$. This is the abelian group given by generators $\vec{x}_0, \vec{x}_1, \ldots, \vec{x}_t$ and relations

$$p_i \vec{x}_i = d_i \vec{x}_0, \quad i = 1, \ldots, t.$$

If all $d_i = 1$, we write $\mathbb{L}(\mathbf{p})$ instead. Note that these groups can have a non-trivial torsion part.

Let $R = k[X_1, X_2]$ be the polynomial algebra graded by total degree. Let $\lambda_1, \ldots, \lambda_t$ be pairwise different elements in $k \cup \{\infty\}$. Without loss of generality we assume $\lambda_1 = \infty$ and $\lambda_2 = 0$. Let $\pi_i \in R$ be the homogeneous prime element $X_2 + \lambda_i X_1$ for $i = 3, \ldots, t$ and $\pi_1 = X_1$, $\pi_2 = X_2$. (In this case, all $d_i = 1$.) Successive insertion of weights $p_i > 1$ into the primes π_i leads to the $\mathbb{L}(\mathbf{p})$-graded algebra

$$k[X_1, X_2, \ldots, X_t]/(X_i^{p_i} - X_2^{p_2} - \lambda_i X_1^{p_1} \mid i = 3, \ldots, t),$$

which are just the projective coordinate algebras of the weighted projective lines described in [**34**].

THEOREM 6.2.4. *Let $\pi \in R$ be a central prime element and $x \in \mathbb{X}$ be the associated point. Let $p \geq 2$ be an integer.*

(1) *$\overline{R} = R[\pi^{1/p}]$ is an $\overline{H}$-graded factorial domain of Krull dimension two. More precisely, the homogeneous prime ideals in $\overline{R}$ of height one are $P = \overline{R}\overline{\pi}$ and $P = \overline{R}q$, where $q \in R$ is prime and not associated to π.*

(2) *There is an equivalence $\frac{\mathrm{mod}^{\overline{H}}(\overline{R})}{\mathrm{mod}_0^{\overline{H}}(\overline{R})} \simeq \mathcal{H}\left(\begin{matrix} p \\ x \end{matrix}\right)$.*

PROOF. (1) We have an embedding $R \subset \overline{R}$, and R can be considered also as $\overline{H}$-graded algebra. $\overline{R} = R[\overline{\pi}]$ is a finite centralizing extension, since $\overline{R} = R \oplus R\overline{\pi} \oplus \cdots \oplus R\overline{\pi}^{p-1}$. Hence, the intersection of a homogeneous prime ideal in $\overline{R}$ with R gives a homogeneous prime ideal in R, proper inclusion is preserved, and each homogeneous prime ideal in R is of this form (see [**77**, 10.]). Consequently, $\overline{R}$ is of graded Krull dimension two.

By the definition of the grading, every homogeneous element $\overline{a} \in \overline{R}$ has the form $\overline{a} = a\overline{\pi}^l$, with $a \in R$ homogeneous and $0 \leq l \leq p - 1$. Hence, $\overline{R}$ is a graded domain like R.

It is easy to see that $\overline{\pi}\overline{R} \cap R = \pi R$, hence there is an isomorphism $R/\pi R \simeq \overline{R}/\overline{\pi}\overline{R}$, and it follows, that $\overline{\pi}$ is a central prime element in $\overline{R}$.

By the form of the homogeneous elements it follows easily that for a homogeneous prime ideal $Rq \subset R$ different from $R\pi$, the ideal $\overline{R}q$ is prime in $\overline{R}$. Moreover, since the map $P \mapsto P \cap R$ preserves proper inclusions, we see that every homogeneous prime ideal of height one in $\overline{R}$ different from $\overline{R}\overline{\pi}$ is of the form $\overline{R}q$, where q is prime in R and not associated to π.

(2) Denote by $\overline{\mathcal{H}}$ the category of p-cycles concentrated in x. Let $\widetilde{}$ and Γ_+ be the functors as defined in 2.1.5. Extending this, we construct an exact functor $\widetilde{} : \mathrm{mod}^{\overline{H}}(\overline{R}) \longrightarrow \overline{\mathcal{H}}$ with kernel $\mathrm{mod}_0^{\overline{H}}(\overline{R})$. Denote by $r : \mathrm{mod}^{\overline{H}}(\overline{R}) \longrightarrow \mathrm{mod}^H(R)$ the exact functor, given by restricting an $\overline{H}$-graded module to the subgroup H. Obviously, $r(\overline{R}) = R$. Moreover, M is of finite length over $\overline{R}$ if and only if $r(M(i\overline{h}))$ is of finite length over R for all $i = 0, \ldots, p - 1$. Hence r induces a functor $\overline{r}$ on the quotient categories. For $M \in \mathrm{mod}^{\overline{H}}(\overline{R})$ and $i = 0, \ldots, p$ define

(6.2.1) $E_i = r(\widetilde{M(i\overline{h})}) \in \mathcal{H}.$

Then $E_p = E_0(x)$ and

$$\widetilde{M} \overset{def}{=} [\, E_0 \xrightarrow{\cdot\overline{\pi}} E_1 \xrightarrow{\cdot\overline{\pi}} E_2 \rightarrow \cdots \rightarrow E_{p-1} \xrightarrow{\cdot\overline{\pi}} E_0(x) \,]$$

defines a p-cycle concentrated in x. In fact, by Theorem 3.1.2 multiplication $(\cdot\overline{\pi})^p = \cdot\pi$ induces the natural transformation $1 \xrightarrow{\varepsilon_x} \sigma_x$. In this way we get the functor $\widetilde{\cdot}$ with the desired properties. By using $\overline{r}$ the induced functor on the quotient category is full, since the sheafification functor is.

It remains to show that $\widetilde{\cdot}$ is dense. Let

$$\overline{E} = [\, E_0 \xrightarrow{x_0} E_1 \xrightarrow{x_1} E_2 \to \cdots \to E_{p-1} \xrightarrow{x_{p-1}} E_0(x)\,]$$

be an arbitrary p-cycle concentrated in x. For each i, define $M_i = \Gamma_+(E_i) \in \mathrm{mod}^H(R)$ and $f_i = \Gamma_+(x_i)$. Define an $\overline{R}$-module by $M = (M_0, \ldots, M_{p-1})$, where the homogeneous elements from $(M_i)_g$ have degrees in $g + i\overline{h}$ (for $g \in H$), and by defining the action of $\overline{\pi}$ on M_i by $m \cdot \overline{\pi} = f_i(m)$. This is well-defined, since multiplication with $\overline{\pi}^p = \pi$ agrees with $f_{i+p-1} \circ \cdots \circ f_i$ (compare 3.1.1). Obviously, $\widetilde{M} \simeq \overline{E}$. $\qquad\square$

REMARK 6.2.5. (1) The proof of the denseness of $\widetilde{\cdot} : \mathrm{mod}^{\overline{H}}(\overline{R}) \longrightarrow \overline{\mathcal{H}}$ shows, that there is a functor $\Gamma_+ : \overline{\mathcal{H}} \longrightarrow \mathrm{mod}^{\overline{H}}(\overline{R})$ such that $\widetilde{\cdot} \circ \Gamma_+$ is isomorphic to the identity functor. Moreover, (restricting to positive gradings) it is right adjoint to $\widetilde{\cdot}$, which follows from the adjointness of the corresponding functors $\widetilde{\cdot} : \mathrm{mod}^{\overline{H_+}}(\overline{R}) \longrightarrow \overline{\mathcal{H}}$ and $\Gamma_+ : \overline{\mathcal{H}} \longrightarrow \mathrm{mod}^{\overline{H_+}}(\overline{R})$ (compare 2.1.5).

(2) Degree shift by $\overline{h}$ on $\mathrm{mod}^{\overline{H}}(\overline{R})$ corresponds to the tubular shift $\overline{\sigma}_x$ on $\overline{\mathcal{H}}$.

(3) Let $y \in \mathbb{X}$, $y \neq x$ be a (homogeneous) point such that the corresponding prime π_y is central. The conclusion in Theorem 3.1.2 (for y) extends (componentwise) to $\overline{\mathcal{H}} = \mathcal{H}\begin{pmatrix} p \\ x \end{pmatrix}$, that is, right multiplication with $\pi_y \in R \subset \overline{R}$ induces the natural transformation $1_{\overline{\mathcal{H}}} \xrightarrow{y} \sigma_y$. $\qquad\square$

COROLLARY 6.2.6. *Let $\pi_1, \ldots, \pi_t \in R$ be central prime elements, which are pairwise non-associated, let $x_1, \ldots, x_t \in \mathbb{X}$ be the associated points, respectively. Denote by $\overline{R}$ the algebra obtained from R by insertion of weights $p_1, \ldots, p_t \geq 2$ into $\pi_1, \ldots, \pi_t$, respectively. Then $\overline{R}$ is a graded factorial algebra and*

$$\frac{\mathrm{mod}^{\overline{H}}(\overline{R})}{\mathrm{mod}_0^{\overline{H}}(\overline{R})} \simeq \overline{\mathcal{H}} = \mathcal{H}\begin{pmatrix} p_1, \ldots, p_t \\ x_1, \ldots, x_t \end{pmatrix}.$$

PROOF. With the preceding remark, we can apply the theorem inductively. $\qquad\square$

REMARK 6.2.7. Let $\overline{R} = R[\pi^{1/p}]$ with $\pi \in R$ a central prime associated to a homogeneous point x.

(1) For $M = \overline{R}$, the E_i in (6.2.1) become $E_i = \widetilde{r(R\overline{\pi}^i)}$ $(i = 0, \ldots, p-1)$, so that $\cdot\overline{\pi}$ defines an isomorphism between E_i and E_{i+1} (for $i < p-1$). Moreover, $E_p = \widetilde{R_0(h)}$ and multiplication with $\overline{\pi}$ gives only a monomorphism $E_{p-1} \longrightarrow E_p$. We see, that the associated p-cycle $\overline{L}$ is isomorphic to

$$j(L) = [\, L = L = L = \cdots = L \xrightarrow{\cdot\pi} L(x)\,],$$

where $j : \mathcal{H} \longrightarrow \overline{\mathcal{H}}$ is the canonical embedding. Moreover, with the tubular shift $\overline{\sigma}_x$ on $\overline{\mathcal{H}}$ we have a morphism of p-cycles

$$
\begin{array}{ccccccccc}
\overline{L} = [& L & = \cdots = & L & = & L & \xrightarrow{\cdot\pi} & L(x) &] \\
& \downarrow & & \| & & \| & & \downarrow{\cdot\pi} & \| \\
\overline{\sigma}_x\overline{L} = [& L & = \cdots = & L & \xrightarrow{\cdot\pi} & L(x) & = & L(x) &]
\end{array}
$$

It is easy to see that $j(L)$ is a special line bundle, if L is.

(2) Assume that $\pi = u_1 \ldots u_e$ with irreducible $u_i \in R$ of degree f. Then the morphism $\overline{L} \longrightarrow \overline{\sigma}_x\overline{L}$ from (1) factorizes into e morphisms between line bundles. For example, for $e = 3$ we get the following picture:

$$
\begin{array}{ccccccccc}
\overline{L} = [& L & = \cdots = & L & = & L & \xrightarrow{\pi\cdot} & L(x) &] \\
& \| & & \| & & \downarrow{u_3\cdot} & & \| & \\
& L & = \cdots = & L & \xrightarrow{u_3\cdot} & L(f) & \xrightarrow{u_1u_2\cdot} & L(x) & \\
& \| & & \| & & \downarrow{u_2\cdot} & & \| & \\
& L & = \cdots = & L & \xrightarrow{u_2u_3\cdot} & L(2f) & \xrightarrow{u_1\cdot} & L(x) & \\
& \| & & \| & & \downarrow{u_1\cdot} & & \| & \\
\overline{\sigma}_x\overline{L} = [& L & = \cdots = & L & \xrightarrow{\pi\cdot} & L(x) & = & L(x) &]
\end{array}
$$

Note that multiplications with the (non-central) u_i from the left act on R as morphisms of right R-modules.

The line bundles lying in between $\overline{L}$ and $\overline{\sigma}_x(\overline{L})$ are not of the form $\widetilde{\overline{R}(g)}$ for some $g \in \overline{H}$. (See also the example 8.5.5.)

(3) The restriction functor $r : \mathrm{mod}^{\overline{H}}(\overline{R}) \longrightarrow \mathrm{mod}^H(R)$, or more precisely, the induced functor $\overline{r}$ between the quotient categories, plays the same role as the functor $r : \overline{\mathcal{H}} \longrightarrow \mathcal{H}$ defined in 6.1.2.   □

With the results of Section 4.3 we get the following.

COROLLARY 6.2.8. *Let $\mathbb{X}$ be an exceptional curve. The following statements are equivalent:*

(1) *$\mathbb{X}$ is commutative.*

(2) *$\mathbb{X}$ is multiplicity free.*

(3) *$\mathbb{X}$ admits a commutative graded factorial domain as projective coordinate algebra.*   □

As a consequence of this section, insertion of weights can be perfectly described on the level of the projective coordinate algebra as far as central prime elements are concerned. One should emphasize that many interesting weighted examples can be examined already in this restricted context. Whereas insertion of weights is established very general by the p-cycle construction, a corresponding description on the level of the graded algebra for weight-insertion into non-central prime elements is still not available.

PROBLEM 6.2.9. Describe the concept of insertion of weights into prime elements which are not central.

6.3. Automorphism groups for weighted curves

Each exceptional curve arises by insertion of weights at finitely many points for some homogeneous exceptional curve [**68**]. The following proposition reduces the problem of calculating the automorphism group essentially to the homogeneous case.

Let $\mathbb{X}$ be a (homogeneous) exceptional curve and $\phi \in \mathrm{Aut}(\mathbb{X})$. Recall that $\overline{\phi}$ denotes the shadow of ϕ. Let $p : \mathbb{X} \longrightarrow \mathbb{N}$ be a weight function. Then ϕ is called *weight preserving* (with respect to p), if $p(\overline{\phi}(x)) = p(x)$ for all $x \in \mathbb{X}$. We have in mind the following situation: Let $x_1, \ldots, x_t \in \mathbb{X}$ be distinguished points and $p_1, \ldots, p_t$ weights. Then let p be defined by $p(x_i) = p_i$ $(i = 1, \ldots, t)$ and $p(x) = 1$ for all x different from the points $x_1, \ldots, x_t$.

PROPOSITION 6.3.1. *Let $\overline{\mathbb{X}}$ be an exceptional curve with underlying homogeneous exceptional curve $\mathbb{X}$ such that $\overline{\mathbb{X}}$ arises from $\mathbb{X}$ by insertion of the weights $p_1, \ldots, p_t$ into the distinct points $x_1, \ldots, x_t$, respectively. Then $\mathrm{Aut}(\overline{\mathbb{X}})$ can be identified with the subgroup of elements in $\mathrm{Aut}(\mathbb{X})$ which preserve these weights.*

PROOF. Let $u \in \mathrm{Aut}(\overline{\mathbb{X}})$. Then there is a unique $v \in \mathrm{Aut}(\mathcal{H})$ such that $\ell u = v\ell$, where $\ell : \overline{\mathcal{H}} \longrightarrow \mathcal{H}$ is left adjoint to the inclusion $j : \mathcal{H} \longrightarrow \overline{\mathcal{H}}$, compare 6.1.2. Moreover, $\ell j \simeq 1_{\mathcal{H}}$. Let L be the structure sheaf of $\mathbb{X}$ as before. Then jL is a special line bundle of $\overline{\mathbb{X}}$. Since $u(jL) = jL$, we get $v(L) \simeq v(\ell jL) = \ell u(jL) = \ell j(L) \simeq L$. If $u \simeq 1_{\overline{\mathcal{H}}}$, then $v \simeq v\ell j \simeq \ell u j \simeq \ell j \simeq 1_{\mathcal{H}}$. It follows that $u \mapsto v$ defines a map $\iota : \mathrm{Aut}(\overline{\mathbb{X}}) \longrightarrow \mathrm{Aut}(\mathbb{X})$, which is a homomorphism of groups.

We show injectivity of this map. Assume, that $v \simeq 1_{\mathcal{H}}$. Then $\ell u \simeq \ell$. By 6.1.4 it follows, that u preserves all simple objects, hence $u \circ \overline{\sigma}_x \simeq \overline{\sigma}_x \circ u$ for each $x \in \overline{\mathbb{X}}$ by 0.4.8. It follows, that u acts like the identity on the components of p-cycles in x and also on the components of morphisms between such cycles. Considering the natural transformation $1 \longrightarrow \overline{\sigma}_x$ it follows that u also acts as the identity on "horizontal" arrows in each cycle. Thus, u acts naturally as identity on p-cycles concentrated in x, that is, $u \simeq 1_{\overline{\mathcal{H}}}$.

Each $v \in \mathrm{Aut}(\mathbb{X})$ lying in the image of ι preserves the weights. Conversely, assume that $v \in \mathrm{Aut}(\mathbb{X})$ preserves the weights $p_1, \ldots, p_t$. Then v can be extended "componentwise" onto cycles, and inductively to an element $u \in \mathrm{Aut}(\overline{\mathbb{X}})$ such that $\ell u = v \ell$. Hence $v = \iota(u)$. (For example, if $\overline{v}(x_1) = x_2$ and $p_1 = p(x_1) = p(x_2) = p_2$, then v can be extended to $u : \mathcal{H}\begin{pmatrix} p_1 \\ x_1 \end{pmatrix} \longrightarrow \mathcal{H}\begin{pmatrix} p_2 \\ x_2 \end{pmatrix}$, and this can be continued.) $\square$

COROLLARY 6.3.2. *Let $\mathbb{X}$ be an exceptional curve. Then the isomorphism class of $\mathrm{Aut}(\mathbb{X})$ is independent of the chosen special line bundle L as structure sheaf.*

PROOF. Let L and L' be two special line bundles. After applying suitable tubular shifts associated to exceptional points, L and L' are special with respect to the same set of exceptional simple objects. By perpendicular calculus L and L' correspond to line bundles on the underlying homogeneous curve $\mathbb{Y}$. But the definition of $\mathrm{Aut}(\mathbb{Y})$ is clearly independent of the choice of the structure sheaf. Then the assertion follows by observing that if an automorphism ϕ of $\mathcal{H}$ fixes L (up to isomorphism), then $\sigma_x \phi \sigma_x^{-1}$ fixes $\sigma_x(L)$ for any exceptional point x. $\square$

Also the calculation of the Picard group reduces to the homogeneous case. Recall the following notion we already used before. Let H be an abelian group,

$h \in H$ and $p \geq 2$ be an integer. Then denote by $H[\frac{h}{p}]$ the abelian group given by $(H \oplus \mathbb{Z})/\mathbb{Z}(-h, p)$. Similarly, $H[\frac{h_1}{p_1}, \dots, \frac{h_t}{p_t}]$ is defined inductively.

The next proposition follows immediately with 6.1.5.

PROPOSITION 6.3.3. *With the same notations as in Proposition 6.3.1, we have*

$$\mathrm{Pic}(\overline{\mathbb{X}}) = \mathrm{Pic}(\mathbb{X})\left[\frac{\sigma_{x_1}}{p_1}, \dots, \frac{\sigma_{x_t}}{p_t}\right]. \qquad \square$$

The following is the extension of Proposition 5.2.3 to the weighted case.

PROPOSITION 6.3.4. *Let $\mathbb{X}$ be an exceptional curve, such that for the underlying homogeneous situation there is an exhaustive automorphism in the Picard group, and such that the underlying bimodule is not of orbit case IIIb. Let $\mathcal{G}$ be the ghost group. Then the group $\mathrm{Pic}(\mathbb{X})/\mathrm{Pic}(\mathbb{X}) \cap \mathcal{G}$ acts simply transitive on the $\mathrm{Aut}(\mathcal{H})$-orbit of the structure sheaf L, and there is a split exact sequence of groups*

$$1 \longrightarrow \mathrm{Pic}(\mathbb{X})/\mathrm{Pic}(\mathbb{X}) \cap \mathcal{G} \longrightarrow \mathrm{Aut}(\mathcal{H})/\mathcal{G} \longrightarrow \mathrm{Aut}(\mathbb{X})/\mathcal{G} \longrightarrow 1.$$

PROOF. Let L' be lying in the same $\mathrm{Aut}(\mathcal{H})$-orbit X as L. After applying suitable shifts associated to exceptional points we can assume that L and L' are special with respect to the same set of exceptional simple objects. By perpendicular calculus, L and L' are line bundles over the associated homogeneous curve. By assumption, there is a Picard element mapping L onto L'. Hence $\mathrm{Pic}(\mathbb{X})$ acts transitively on X. Each ghost fixes L, hence also any other member of X. We get an induced action of $\mathrm{Pic}(\mathbb{X})/\mathrm{Pic}(\mathbb{X}) \cap \mathcal{G}$ on X, which is obviously simply transitive.

Define $\mathrm{Aut}(\mathcal{H})/\mathcal{G} \longrightarrow \mathrm{Aut}(\mathbb{X})/\mathcal{G}$ by $[\phi] \mapsto [\sigma \circ \phi]$, where $\sigma \in \mathrm{Pic}(\mathbb{X})$ such that $\sigma(\phi(L)) \simeq L$. This induces the split exact sequence. $\qquad \square$

In the special situation $k = \mathbb{R}$ Proposition 6.3.4 can be formulated as follows.

PROPOSITION 6.3.5. *Let $\mathbb{X}$ be an exceptional curve over the real numbers. Let $\mathcal{G}$ be the ghost group. Then the group $\mathrm{Pic}(\mathbb{X})/\mathcal{G}$ acts simply transitive on the set of all (isomorphism classes of) special line bundles and there is a split exact sequence of groups*

$$1 \longrightarrow \mathrm{Pic}(\mathbb{X})/\mathcal{G} \longrightarrow \mathrm{Aut}(\mathcal{H})/\mathcal{G} \longrightarrow \mathrm{Aut}(\mathbb{X})/\mathcal{G} \longrightarrow 1. \qquad \square$$

Note that [**58**, Lem. 4+Thm. 5] is not quite correct in the twisted case $\mathbb{C} \oplus \overline{\mathbb{C}}$, where $\mathcal{G}$ is non-trivial; moreover, we have to restrict to *special* line bundles as in the preceding proposition. We will give an example, where there are line bundles which are not special in 8.5.1.

For the domestic curves and the tubular curves over the real numbers the automorphism groups are listed in Appendix A.1.

CHAPTER 7

Exceptional objects

In this chapter we briefly expose two examples of problems in the context of exceptional objects. The first is the proof of the transitivity of the braid group action on the set of complete exceptional sequences over an exceptional curve which shows that the result is independent of the base field's arithmetic. By contrast, the second example does not carry over to an arbitrary field. It deals with the characterization of exceptional curves by graded factoriality.

7.1. Transitivity of the braid group action

In this section we report on a joint result with H. Meltzer [60] which supports the philosophy that results on exceptional objects are essentially independent from the base field.

Let $\mathbb{X}$ be an exceptional curve with hereditary category $\mathcal{H}$. A sequence $(E_1, \ldots, E_n)$ of exceptional objects in $\mathcal{H}$ is called exceptional sequence, if for all $j > i$ we have $\mathrm{Hom}(E_j, E_i) = 0 = \mathrm{Ext}^1(E_j, E_i)$. It is called complete, if n coincides with the rank of the Grothendieck group of $\mathcal{H}$.

The notion of complete exceptional sequences is closely related to the concept of tilting objects (complexes). We remark that there is a characterization of exceptional curves similar to 0.3.6 in terms of the existence of a complete exceptional sequence instead of a tilting object [68].

The braid group B_n on n strands is defined by generators $\sigma_1, \ldots, \sigma_{n-1}$ and relations $\sigma_i \sigma_j = \sigma_j \sigma_i$ for $j \geq i + 2$ and $\sigma_i \sigma_{i+1} \sigma_i = \sigma_{i+1} \sigma_i \sigma_{i+1}$ for $i = 1, \ldots, n - 2$.

The braid group B_n acts on the set of all exceptional sequences of length n: σ_i replaces in $(E_1, \ldots, E_n)$ the pair (E_i, E_{i+1}) by the pair $(E_{i+1}, R_{E_{i+1}}(E_i))$, where $R_{E_{i+1}}(E_i)$ is the right mutation of E_i by E_{i+1} (see [60] for details).

THEOREM 7.1.1 ([60]). *Let $\mathbb{X}$ be an exceptional curve and let n be the rank of the Grothendieck group of $\mathbb{X}$. Then the braid group B_n acts transitively on the set of complete exceptional sequences in $\mathcal{H}$.*

Exceptional vector bundles play an important role in algebraic geometry in the study of vector bundles over various projective varieties, and were introduced by Drezet and Le Potier [31] in connection with the investigation of stable bundles. Exceptional sequences and the braid group action were studied by Bondal [10]. Transitivity of this action on the set of complete exceptional sequences was shown for $\mathbb{P}^2$ by Rudakov [98], for $\mathbb{P}^1 \times \mathbb{P}^1$ by Rudakov [97] and for arbitrary del Pezzo surfaces by Kuleshov and Orlov [52]. We remark that there is a related concept of (collections of) spherical objects in the context of mirror symmetry, see [101].

In representation theory the importance of exceptional objects is without question. The transitivity of the braid group action on the set of complete exceptional sequences was established over an algebraically closed field for the category

of modules over a hereditary algebra by Crawley-Boevey [20] and for the category
of coherent sheaves over a weighted projective line by Meltzer [78]. Ringel [94]
simplified and extended Crawley-Boevey's result to arbitrary hereditary Artin algebras. This last result gave a hint that Meltzer's result should also be true for
arbitrary exceptional curves. On the other hand, the proof for an algebraically
closed field presented in [78] did not work over an arbitrary field. Moreover, the
results in [53, 56] on tubular curves have shown that it is often not predictable
whether new effects occur or not.

We now briefly sketch the idea of the proof from [60]. The proof is by induction
on the rank n of the Grothendieck group $\mathrm{K}_0(\mathbb{X})$. The following lemma is crucial.

LEMMA 7.1.2. *Let $\mathbb{X}$ be a non-homogeneous exceptional curve. Then each complete exceptional sequence in $\mathcal{H}$ can be shifted by the braid group to an exceptional
sequence which contains a simple object.*

7.1.3 (Proof of Theorem 7.1.1). Relying on the lemma, the proof of Theorem 7.1.1 is straight-forward by induction, like in the algebraically closed case: For
$n = 2$, that is, if $\mathbb{X}$ is homogeneous, the proof is easy. Assume $n > 2$. Then there is a
"canonical" complete exceptional sequence $\mathbf{C} = (C_1, \ldots, C_n)$ given by a certain tilting bundle in $\mathcal{H}$ ([70]). Given any complete exceptional sequence $\mathbf{E} = (E_1, \ldots, E_n)$
one can assume by the lemma that $E_n = S$ is an exceptional simple object. By
the special structure of $\mathbf{C}$ we can also assume $C_n = S$. Then we consider the right
perpendicular category $S^\perp$ of S, which is an exceptional curve where the rank of
the Grothendieck group is $n - 1$, and use the induction hypothesis.   □

The proof of Lemma 7.1.2 uses the following rank formula [60] which follows
from [43]. Forming the right perpendicular category $E^\perp$ to an exceptional vector
bundle E, we switch to a module category $E^\perp \simeq \mathrm{mod}(\Lambda)$, where Λ is hereditary
(not necessarily connected) with $n - 1$ simple modules.

PROPOSITION 7.1.4. *Let E be an exceptional object in $\mathcal{H}_+$. Denote by
$S_1, \ldots, S_{n-1}$ a complete system of simple modules in $E^\perp$ and by $P_1, \ldots, P_{n-1}$ their
projective covers. Then*

$$\frac{\mathrm{rk}(E)^2}{[\mathrm{End}(E):k]} = \sum_{i=1}^{n-1} \frac{\mathrm{rk}(P_i) \cdot \mathrm{rk}(S_i)}{[\mathrm{End}(P_i):k]}.$$

7.1.5 (Proof of Lemma 7.1.2). It is sufficient to show that in each orbit $\mathcal{O}$
there is a complete exceptional sequence $\mathbf{E} = (E_1, \ldots, E_n)$ such that E_i is of finite
length for some i. Assume that for the orbit $\mathcal{O}$ this is not the case. Then in $\mathcal{O}$
appears an exceptional vector bundle E, such that $\frac{\mathrm{rk}(E)^2}{[\mathrm{End}(E):k]}$ is minimal. We can
assume, that E appears in $\mathbf{E}$ with $E = E_n$. Forming $E^\perp$, the exceptional sequence
$(E_1, \ldots, E_{n-1})$ is complete in $E^\perp$. Using the transitivity of the braid group action
for $\mathrm{mod}(\Lambda)$ proved by Ringel [94], the sequence $\mathbf{E}$ can be shifted to the exceptional
sequences $(P_1, \ldots, P_{n-1}, E)$ and $(S_{n-1}, \ldots, S_1, E)$, where the P_i and the S_i are as
in the preceding proposition, suitably ordered with $P_1 = S_1$. From the rank formula
we get the contradiction $\frac{\mathrm{rk}(P_1)^2}{[\mathrm{End}(P_1):k]} < \frac{\mathrm{rk}(E)^2}{[\mathrm{End}(E):k]}$.   □

Over an arbitrary field k the endomorphism ring of an exceptional object is a
finite dimensional skew field over k and need not to coincide with k itself. Therefore
the following corollary is an important consequence of the transitivity of the braid
group action.

COROLLARY 7.1.6. *The list of endomorphism skew fields appearing in a complete exceptional sequence in $\mathcal{H}$ is invariant.* □

7.2. Exceptional objects and graded factoriality

If k is algebraically closed then there is a relationship between the concept of graded factoriality and the existence of exceptional objects, as illustrated by the following results:

7.2.1. For a smooth projective curve C over an algebraically closed field k, the following are equivalent [69]:

(1) C is of genus zero.
(2) $\mathrm{coh}(C)$ admits an exceptional object.
(3) $\mathrm{coh}(C)$ admits a tilting object.
(4) There is a (commutative) $\mathbb{Z}$-graded factorial k-algebra R, affine of Krull dimension two, such that $\mathrm{coh}(C) \simeq \frac{\mathrm{mod}^{\mathbb{Z}}(R)}{\mathrm{mod}_0^{\mathbb{Z}}(R)}$.

Moreover, it follows from [53] that this is also true for $k = \mathbb{R}$.

7.2.2. A similar result which follows from [55, 67], see [68] is: Let $\mathcal{H}$ be a small abelian connected category over an algebraically closed field k. Then the following assertions are equivalent:

(1) $\mathcal{H}$ is equivalent to the category of coherent sheaves over an exceptional curve.
(2) $\mathcal{H}$ is of the form $\frac{\mathrm{mod}^H(R)}{\mathrm{mod}_0^H(R)}$ for a (commutative) H-graded factorial affine k-algebra R of Krull dimension two, where H is a finitely generated abelian group of rank one.

The results of Chapters 1 and 6 indicate that the implication (1)⇒(2) (replacing "commutative" by "noncommutative") remains valid for an arbitrary base field (up to the insertion of weights into non-central prime elements). But the converse and also 7.2.1 is wrong in general, even in a commutative situation, as Lenzing pointed out in [69]:

EXAMPLE 7.2.3. Let $k = \mathbb{F}_2$ and R be the commutative $\mathbb{Z}$-graded algebra
$$\mathbb{F}_2[X, Y, Z]/(X^6 + Y^3 + Z^2 + X^2 Y^2 + X^3 Z),$$
where $\deg(X) = 1$, $\deg(Y) = 2$ and $\deg(Z) = 3$. (Note that R is not generated in degree zero and one.) Then R is $\mathbb{Z}$-graded factorial and the quotient category $\frac{\mathrm{mod}^{\mathbb{Z}}(R)}{\mathrm{mod}_0^{\mathbb{Z}}(R)}$ is equivalent to the category $\mathrm{coh}(C)$ of coherent sheaves of a smooth projective curve C of genus one (and not zero). □

It would be interesting to characterize the class of (noncommutative) graded factorial algebras which are related to the exceptional curves.

CHAPTER 8

Tubular exceptional curves

Let Σ be a concealed canonical algebra over a field k with corresponding exceptional curve $\mathbb{X}$. Σ (and $\mathbb{X}$) are called *tubular* if the radical of the Grothendieck group $K_0(\Sigma)$ is finitely generated abelian of rank two. Equivalently, the Coxeter transformation is of finite order. Note that the radical for non-tubular Σ ($\mathbb{X}$, respectively) has always rank one.

An exceptional curve $\mathbb{X}$ is tubular if and only if its virtual genus [66, 68, 70]

$$g_\mathbb{X} = 1 + \frac{\varepsilon p}{2} \left[\sum_{i=1}^{t} d_i \left(1 - \frac{1}{p_i} \right) - \frac{2}{\varepsilon} \right]$$

is one. (Here, p is the least common multiple of the weights $p_1, \ldots, p_t$, compare also 0.4.5.) From this property, it is not surprising that tubular exceptional curves have a strong affinity to elliptic curves $\mathbb{T}$. In both cases, $\mathrm{coh}(\mathbb{X})$ and $\mathrm{coh}(\mathbb{T})$, all indecomposable objects lie in tubes (in the language of representation theory).

More precisely, Atiyah's classification [4] of vector bundles over an elliptic curve $\mathbb{T}$ over an algebraically closed field k shows that $\mathrm{coh}(\mathbb{T})$ consists of a rational family of tubular families, each parametrized by $\mathbb{T}$ and consisting of homogeneous tubes. (Note, that here "rational family" means "indexed by the rational numbers".)

In [91] Ringel introduced the tubular (canonical) algebras over an algebraically closed field and showed that the indecomposable modules over such an algebra can be classified basically by a rational family of tubular families, each parametrized by the projective line $\mathbb{P}^1(k)$; in each tubular family there are finitely many non-homogeneous tubes. Accordingly, the (bounded) derived category (see [37]) of a tubular algebra consists entirely of tubes [42].

The connection between these geometric and representation theoretic results was given by Geigle and Lenzing when they introduced the weighted projective lines [34] and later by Lenzing and Meltzer for the tubular case [71]. The fundamental concept there was that of tubular mutations [79], which are automorphisms of the derived category. (In this tubular situation the tubular *shifts* form a very small subclass of them.)

An additional feature of the mentioned results over an algebraically closed field is that for a fixed tubular algebra (tubular exceptional curve, elliptic curve, respectively) all tubular families are equivalent categories. We showed in [53, 56] that this is no longer true over an arbitrary field. There are tubular exceptional curves which admit tubular families which are not equivalent. Accordingly, there are tubular exceptional curves $\mathbb{X}$ and $\mathbb{X}'$ which are derived equivalent but not isomorphic. In different terminology, $\mathbb{X}$ and $\mathbb{X}'$ are Fourier-Mukai partners.

The present chapter is devoted to the study of the automorphism group $\mathrm{Aut}(\mathrm{D}^b(\mathbb{X}))$ of the bounded derived category of a tubular exceptional curve $\mathbb{X}$. This group acts on the set of all separating tubular families in $\mathrm{D}^b(\mathbb{X})$. Over an algebraically closed field the preceding remark implies that this group action is transitive, but over an arbitrary field there may occur more than one orbit. The number of these orbits is called the index of $\mathbb{X}$. Our main result in [**59**] is that the index of a tubular exceptional curve $\mathbb{X}$ is at most three and that such curves of index three exist. We summarize a proof for this result and present the new Proposition 8.1.6 which improves the argument.

We study examples exploiting the results from the previous chapters. They illustrate the principle how to determine the automorphism group $\operatorname{Aut}(\mathrm{D}^b(\mathbb{X}))$ in general. The central example will be a tubular exceptional curve of index three. In this example, the Grothendieck group is of rank three. (In general, the rank of the Grothendieck group of a tubular exceptional curve is at least three and at most ten [**66**].) Tubular exceptional curves with this property are of particular interest. First of all, since there is only one exceptional tube (of rank two) in each tubular family, exceptional objects are essentially determined by their slope and explicit calculations are much easier than for other tubular curves. This was demonstrated by Ringel [**95**], pointing out an interesting link between tilting modules and Farey fractions.

Moreover, in the tubular case the following effects arise only when $\mathrm{K}_0(\mathbb{X})$ is of rank three:

- the occurrence of index three;
- the occurrence of roots (even 1-roots) in $\mathrm{K}_0(\mathbb{X})$ which are not realizable by indecomposable objects in $\mathcal{H}$ (we refer to [**53, 59**]).

In the algebraically closed case each line bundle L over an exceptional curve $\mathbb{X}$ is exceptional. Over an arbitrary field this is also true for line bundles over a domestic exceptional curve (that is, when the virtual genus satisfies $g_\mathbb{X} < 1$). We will show that it is false for some tubular cases where the Grothendieck group is of rank three or four.

8.1. Slope categories and the rational helix

Throughout this section let $\mathbb{X}$ be a tubular exceptional curve over a field k.

8.1.1 (Slope). For each $\mathbf{x} \in \mathrm{K}_0(\mathbb{X})$ such that $\operatorname{rk}\mathbf{x} \neq 0$ or $\deg\mathbf{x} \neq 0$ define the *slope* by $\mu\mathbf{x} = \frac{\deg\mathbf{x}}{\operatorname{rk}\mathbf{x}}$. The slope of a non-zero object in $\mathcal{H}$ is defined as the slope of its class. *Stability* and *semistability* of non-zero objects in $\mathcal{H}$ is defined with respect to the slope as in [**34**]. For each $q \in \widehat{\mathbb{Q}} := \mathbb{Q} \cup \{\infty\}$ denote by $\mathcal{H}^{(q)}$ the full subcategory of $\mathcal{H}$ which is formed by the zero sheaf and the semistable sheaves of slope q. We call the categories $\mathcal{H}^{(q)}$ (and also their translates in the derived category) *slope categories*. Note that for example $\mathcal{H}^{(\infty)} = \mathcal{H}_0$.

8.1.2. Since each indecomposable object in $\mathcal{H}$ is semistable (compare [**34**, 5.5]), $\mathcal{H}$ is the additive closure of its slope categories, and since $\mathcal{H}$ is hereditary we have

$$\mathcal{D} := \mathrm{D}^b(\mathbb{X}) = \bigvee_{n\in\mathbb{Z}} \mathcal{H}[n] = \bigvee_{(n,q)\in\mathbb{Z}\times\widehat{\mathbb{Q}}} \mathcal{H}^{(q)}[n],$$

which means that $\mathrm{D}^b(\mathbb{X})$ is the additive closure of the (disjoint) copies $\mathcal{H}[n]$ and also of the $\mathcal{H}^{(q)}[n]$, and moreover, there are non-zero morphisms from $\mathcal{H}[n]$ to $\mathcal{H}[n']$ (from $\mathcal{H}^{(q)}[n]$ to $\mathcal{H}^{(q')}[n']$, resp.) only if $n \leq n'$ $((n,q) \leq (n',q')$, resp., where the *rational helix* $\mathbb{Z}\times\widehat{\mathbb{Q}}$ is endowed with the lexicographical order [**72**]). More precisely, for all X, $Y \in \mathcal{H}$ and all m, $n \in \mathbb{Z}$ we have $\mathrm{Hom}_{\mathcal{D}}(X[n], Y[m]) = \mathrm{Ext}_{\mathcal{H}}^{m-n}(X, Y)$. Note that $\mathrm{Ext}_{\mathcal{H}}^{i}(-, -) = 0$ for $i \in \mathbb{Z}$, $i \neq 0$, 1. Here, $X[n]$ denotes the element in the copy $\mathcal{H}[n]$ which corresponds to $X \in \mathcal{H}$. The automorphism T on $\mathcal{D}$, which is induced by the assignment $X \mapsto X[1]$, is called *translation functor*.

8.1.3 (Riemann-Roch formula). Let p be the least common multiple of the weights $p_1, \ldots, p_t$. Recall that for any $\mathbf{x}$, $\mathbf{y} \in \mathrm{K}_0(\mathbb{X})$

$$\langle\!\langle \mathbf{x}, \mathbf{y} \rangle\!\rangle = \sum_{j=0}^{p-1} \langle \tau^j \mathbf{x}, \mathbf{y} \rangle.$$

For any $\mathbf{x}$, $\mathbf{y} \in \mathrm{K}_0(\mathbb{X})$ the following formula holds ([**66, 70**]).

$$\langle\!\langle \mathbf{x}, \mathbf{y} \rangle\!\rangle = \kappa\varepsilon \begin{vmatrix} \mathrm{rk}\,\mathbf{x} & \mathrm{rk}\,\mathbf{y} \\ \deg\mathbf{x} & \deg\mathbf{y} \end{vmatrix},$$

which in case $\mathrm{rk}\,\mathbf{x} \neq 0 \neq \mathrm{rk}\,\mathbf{y}$ can also be written as $\langle\!\langle \mathbf{x}, \mathbf{y} \rangle\!\rangle = \kappa\varepsilon\,\mathrm{rk}\,\mathbf{x}\,\mathrm{rk}\,\mathbf{y}(\mu\mathbf{y} - \mu\mathbf{x})$. As application one gets: If X, $Y \in \mathcal{H}$ are indecomposable with $\mu(X) < \mu(Y)$, then $\mathrm{Hom}(X, \tau^j Y) \neq 0$ for some j.

PROBLEM 8.1.4 (Calabi-Yau property). If k is algebraically closed then the Auslander-Reiten translation, that is, the Serre functor τ on $\mathcal{H}$ is of order p in the group $\mathrm{Aut}(\mathcal{H})$ (where p is the least common multiple of the weights). By Serre duality we conclude that the triangulated category $\mathrm{D}^b(\mathbb{X})$ is Calabi-Yau of fractional dimension p/p, in the sense of [**48**].

It is an interesting question whether this is also always true if k is an arbitrary field. A priori on has to take ghosts into account.

8.1.5 (Interval categories). For each $q \in \overline{\mathbb{Q}}$ let $\mathcal{H}\langle q \rangle$ be the subcategory in $\mathcal{D}$ defined by

$$\mathcal{H}\langle q \rangle = \mathcal{H}_-^{(q)}[-1] \vee \mathcal{H}_+^{(q)} \vee \mathcal{H}^{(q)},$$

where

$$\mathcal{H}_+^{(q)} = \bigvee_{-\infty < r < q} \mathcal{H}^{(r)}, \quad \mathcal{H}_-^{(q)} = \bigvee_{q < r \leq \infty} \mathcal{H}^{(r)}.$$

Moreover

$$\mathcal{D} = \mathcal{D}_+^{(q)} \vee \mathcal{H}^{(q)} \vee \mathcal{D}_-^{(q)},$$

where $\mathcal{D}_+^{(q)} = \{X \in \mathcal{D} \mid \mathrm{Hom}(\mathcal{H}^{(q)}, X) = 0\}$ and $\mathcal{D}_-^{(q)} = \{Y \in \mathcal{D} \mid \mathrm{Hom}(Y, \mathcal{H}^{(q)}) = 0\}$. We call the categories $\mathcal{H}\langle q \rangle$ and also their translates in $\mathcal{D}$ *interval categories*. $\square$

The first proof of the following proposition was a by-product in [**56, 59**] of the case by case analysis how the automorphism group $\mathrm{Aut}(\mathcal{D})$ acts on the set of slope categories. We now give a more systematic argument.

PROPOSITION 8.1.6. *Let $\mathbb{X}$ be a tubular exceptional curve. For each $q \in \widehat{\mathbb{Q}}$ the interval category $\mathcal{H}\langle q \rangle$ is the sheaf category of a tubular exceptional curve $\mathbb{X}\langle q \rangle$.*

PROOF. The key-point (see [**56**, Prop. 7]) is to show that the slope category $\mathcal{H}^{(q)}$ is non-trivial. We modify the argument given in [**56**]. There is a normalized rank function rk_q on $\mathrm{K}_0(\mathbb{X})$ such that $\mathrm{rk}_q(F) \geq 0$ for all objects $F \in \mathcal{H}\langle q \rangle$ and $\mathrm{rk}_q(F) = 0$ if and only if $F \in \mathcal{H}^{(q)}$. In fact, if $q = d/r$, where d and r are coprime, then (up to normalization) $\mathrm{rk}_q(F) = d \cdot \mathrm{rk}(F) - r \cdot \deg(F)$. Moreover, by semistability there is no non-zero morphism from an object of rank zero to an object of non-zero rank. Since $\mathcal{H}^{(q)}$ is noetherian (possibly trivial), noetherianness of the category $\mathcal{H}\langle q\rangle$ follows straight-forwardly, see [**74**, Lem. 5.2].

Let $\mathcal{H}\langle q\rangle_0$ be the subcategory in $\mathcal{H}\langle q\rangle$ of objects of finite length which is a Serre subcategory. By [**74**] the quotient category $\mathcal{H}\langle q\rangle/\mathcal{H}\langle q\rangle_0$ is a length category, and its length function defines a rank function on $\mathcal{H}\langle q\rangle$. This rank function is (up to some normalization factor) of the form $\mathrm{rk}_{q'}$ for some slope q' (see [**57**, Prop. 5.3]). It follows, that $q' \leq q$ and (up to translation) $\mathcal{H}\langle q\rangle_0 = \mathcal{H}^{(q')}$, which is non-trivial by noetherianness of $\mathcal{H}\langle q\rangle$. Assume that $q' < q$. Then:

(i) By semistability and noetherianness there is no non-zero object of slope r with $q' < r \leq q$. (If $0 \neq F$ has slope r consider a maximal subobject of F; the simple factor then lies in $\mathcal{H}^{(q')}$ which gives a non-zero map from F to an object of smaller slope.)

(ii) Since the category $\mathcal{H}\langle q'\rangle$ is connected there is a non-zero torsion-free object in $\mathcal{H}\langle q'\rangle$. Shifting this object sufficiently far to the left leads to a non-zero object of slope r with $q' < r \leq q$, a contradiction to (i).

Therefore $q = q'$ follows, hence $\mathcal{H}^{(q)}$ is also non-trivial. $\qquad\square$

It follows that the slope induces a bijection between all slope categories in $\mathrm{D}^b(\mathbb{X})$ and the elements of the *rational helix* $\mathbb{Y} = \mathbb{Z} \times \widehat{\mathbb{Q}}$.

8.1.7. Moreover, it follows that for all $q \in \widehat{\mathbb{Q}}$ there is defined the *q-symbol*, that is, the symbol of the curve $\mathbb{X}\langle q\rangle$.

For all q, $q' \in \widehat{\mathbb{Q}}$ the tubular exceptional curves $\mathbb{X}\langle q\rangle$ and $\mathbb{X}\langle q'\rangle$ are derived equivalent, that is, they are Fourier-Mukai partners. In particular, all the Grothendieck groups $\mathrm{K}_0(\mathbb{X}\langle q\rangle)$ and $\mathrm{K}_0(\mathbb{X}\langle q'\rangle)$ (equipped with the Euler forms) are isomorphic. This means, by definition, that the symbols of $\mathbb{X}\langle q\rangle$ and $\mathbb{X}\langle q'\rangle$ are equivalent, but in general they are different, and accordingly the curves $\mathbb{X}\langle q\rangle$ and $\mathbb{X}\langle q'\rangle$ non-isomorphic, that is, $\mathcal{H}\langle q\rangle \not\simeq \mathcal{H}\langle q'\rangle$. (Compare the list of (equivalence classes of) tubular symbols given in Appendix B.)

Furthermore, for all q the tubular shifts associated to points in $\mathbb{X}\langle q\rangle$ are defined and are automorphisms of $\mathcal{H}\langle q\rangle$, which extend to elements in $\mathrm{Aut}(\mathcal{D})$. These are by definition the tubular mutations.

LEMMA 8.1.8 ([**56**, Cor. 11]). *Let $\phi \in \mathrm{Aut}(\mathcal{D})$. For any element (n, q) of the rational helix there is a unique (n', q') in the rational helix such that $\phi(\mathcal{H}^{(q)}[n]) = \mathcal{H}^{(q')}[n']$. Hence, by setting $\overline{\phi}(n, q) = (n', q')$ we get an automorphism $\overline{\phi}$ of the rational helix. This induces a homomorphism of groups $\Phi : \mathrm{Aut}(\mathcal{D}) \longrightarrow \mathrm{Aut}(\mathbb{Y})$, $\phi \mapsto \overline{\phi}$.* $\qquad\square$

8.1.9. Note that $\mathrm{Aut}(\mathbb{Y}) \simeq B_3$, the braid group on three strands, which is defined by generators s, ℓ and the relation $s\ell s = \ell s\ell$ [**72**]. The translation T is mapped under Φ to $t = (s\ell)^3$, which is a central element of infinite order. We have the exact sequence

$$1 \longrightarrow \langle t\rangle \longrightarrow B_3 \overset{p}{\longrightarrow} \mathrm{PSL}_2(\mathbb{Z}) \longrightarrow 1,$$

given by $\ell \mapsto \begin{pmatrix} 1 & 0 \\ -1 & 1 \end{pmatrix}$, $s \mapsto \begin{pmatrix} 1 & 1 \\ 0 & 1 \end{pmatrix}$.

We described in [**59**] which subgroups of B_3 occur as images of Φ in the different cases (compare also [**57**, Table 1]). Typically the image of Φ is generated by the images of two or three tubular mutations and the translation in the derived category. Here we write "typically", since the situation is not fully clarified in the cases where the rank of the Grothendieck group is three. See also Remark 8.2.9 below.

8.1.10. The kernel of Φ is given by the automorphisms preserving the slope. Since these automorphisms restrict naturally to $\mathcal{H}$, we consider them as elements in $\mathrm{Aut}(\mathcal{H})$, which defines the subgroup $\mathrm{Aut}_\mu(\mathcal{H})$. With the assumptions of Proposition 6.3.4 the slope preserving automorphisms are just those of $\mathrm{Pic}_0(\mathbb{X})$ and of $\mathrm{Aut}(\mathbb{X})$ (and compositions of them), and with the ghost group $\mathcal{G}$ we have

$$\mathrm{Aut}_\mu(\mathcal{H})/\mathcal{G} \simeq \left(\mathrm{Pic}_0(\mathbb{X})/\mathrm{Pic}_0(\mathbb{X}) \cap \mathcal{G} \right) \rtimes \mathrm{Aut}(\mathbb{X})/\mathcal{G}.$$

In particular it follows, that if $\mathbb{X}$ and $\mathbb{X}'$ are derived equivalent tubular exceptional curves, then there is a relationship between $\mathrm{Pic}_0(\mathbb{X})$ and $\mathrm{Aut}(\mathbb{X})$ on the one hand and $\mathrm{Pic}_0(\mathbb{X}')$ and $\mathrm{Aut}(\mathbb{X}')$ on the other hand. But it is not true in general that the automorphism groups of $\mathbb{X}$ and $\mathbb{X}'$ are isomorphic, neither the geometric automorphism groups nor the ghost groups (compare 8.3.2).

8.2. The index of a tubular exceptional curve

We keep the notations from the previous section.

If k is algebraically closed the map Φ is always surjective. Equivalently, $\mathrm{Aut}(\mathcal{D})$ acts transitively on $\mathbb{Y}$ in this case. But in general Φ is not surjective.

DEFINITION 8.2.1. Let $\mathbb{X}$ be a tubular exceptional curve. Then the number of $\mathrm{Aut}(\mathrm{D}^b(\mathbb{X}))$-orbits in the rational helix $\mathbb{Y}$ is called the *index* of $\mathbb{X}$. Similarly, the index of a tubular algebra is defined.

THEOREM 8.2.2 ([**53, 59**]). *The index of a tubular exceptional curve is at most three.*

In the following we sketch the idea of the proof. For details we refer to [**59**].

8.2.3. Denote by $V = \mathrm{K}_0(\mathbb{X})$ the Grothendieck group of $\mathcal{H}$, by $R = \mathrm{Rad}(V)$ the radical of V, and by $\mathbb{P}R$ the set of direct summands of R of rank one. Taking the slope of generators of such direct summands induces a bijection between $\mathbb{P}R$ and $\widehat{\mathbb{Q}}$ (see [**57**]). There is the following commutative diagram

$$
\begin{array}{ccc}
\mathrm{Aut}(\mathcal{D}) & \xrightarrow{\ \Phi\ } & \mathrm{Aut}(\mathbb{Y}) = B_3 \\
{\scriptstyle \kappa}\downarrow & & \downarrow{\scriptstyle p} \\
\mathrm{Aut}(V) & \xrightarrow{\ \Psi\ } & \mathrm{Aut}(\mathbb{P}R) = \mathrm{PSL}_2(\mathbb{Z}).
\end{array}
$$

Each element in $\mathrm{Aut}(V)$ induces by restriction to the radical an automorphism of $\mathbb{P}R$, which defines the map Ψ. The automorphism group of $\mathbb{P}R$ can be identified with the projective modular group $\mathrm{PSL}_2(\mathbb{Z})$ (see [**57**]). κ is defined by $\kappa(\phi)([X]) = [\phi X]$ for any $\phi \in \mathrm{Aut}(\mathcal{D})$ and any $X \in \mathcal{D}$.

8.2.4. It is shown in [**57**] that the group $\mathrm{Aut}(V)$ acts on $\widehat{\mathbb{Q}}$ with at most two orbits. This is shown by determining a subgroup of $\mathrm{Aut}(V)$ generated by a few certain shift automorphisms (defined on the K-theoretical level, see [**66**]) such that the Ψ-image of this subgroup coincides with the Ψ-image of $\mathrm{Aut}(V)$ and therefore this subgroup acts with at most two orbits. Then the idea of the proof of

Theorem 8.2.2 is to realize these K-theoretical automorphisms by tubular mutations on the derived level. In the cases where the rank of the Grothendieck group is greater or equal than four this can be accomplished without problem, so that there are also at most two orbits on the derived level. But if the rank of the Grothendieck group equals three (that is, in each tubular family there is precisely one exceptional tube, and this is of rank two) the index depends also on the arithmetic of the base field, which leads to cases of index three (see Proposition 8.2.5 below; see also Appendix B for a list of the 17 tubular cases). Moreover, the analysis in [**59**] shows the following propositions which stress the special role of the rank three case.

PROPOSITION 8.2.5. *Let* $\mathbb{X}$ *be a tubular exceptional curve such that the Grothendieck group is of rank three. Then the index of* $\mathbb{X}$ *is at most three.*

Assume that there exists $q \in \widehat{\mathbb{Q}}$ *such that the numerical type of* $\mathbb{X}\langle q \rangle$ *is* $\varepsilon = 1$ *and such that there exists a unirational point in* $\mathbb{X}\langle q \rangle$. *Then the index of* $\mathbb{X}$ *is at most two.*

If the symbol of $\mathbb{X}$ *equals* $\begin{pmatrix} 2 \\ 4 \\ 2 \end{pmatrix}$ *and if, for example,* $\mathbb{X}$ *and* $\mathbb{X}\langle 2 \rangle$ *contain unirational points, then the index is one.*

PROOF. Let $q \in \widehat{\mathbb{Q}}$ such that the numerical type of $\mathbb{X}\langle q \rangle$ is $\varepsilon = 1$. Switching to $\mathbb{X}\langle q \rangle$ we can assume that the numerical type of $\mathbb{X}$ itself is $\varepsilon = 1$. The Φ-image of the tubular mutations with respect to the exceptional tubes containing the structure sheaf L and an exceptional simple object together with the Φ-image of the translation T gives the subgroup $\langle t, \ell, s^4 \rangle$ of B_3. This subgroup acts with three orbits on the rational helix. The remaining assertions follow from the analysis in [**57**, 10.]. $\qquad\square$

It is easy to see that each $(2,2)$-bimodule over a finite field k is non-simple, hence there is a unirational point for the associated curve (compare 0.6.2). The same is true for the field $k = \mathbb{R}$ of real numbers. Thus we get

COROLLARY 8.2.6. *The index of a tubular algebra (or tubular exceptional curve) over any finite field* k, *or over the field* $\mathbb{R}$ *of real numbers, is at most two.* $\qquad\square$

PROPOSITION 8.2.7 ([**59**]). *Let* $\mathbb{X}$ *be a tubular exceptional curve and* r *be the rank of its Grothendieck group. There is a subgroup* U *of* $\mathrm{Aut}(\mathrm{D}^b(\mathbb{X}))$ *acting transitively on each* $\mathrm{Aut}(\mathrm{D}^b(\mathbb{X}))$-*orbit in the set of all slope categories such that additionally the following holds:*

If $r \geq 4$, *or if the index is three, then* U *is generated by the translation* T *and two tubular mutations associated to exceptional tubes.* $\qquad\square$

PROPOSITION 8.2.8. *Let* $\mathbb{X}$ *be a tubular exceptional curve such that the Grothendieck group is of rank three.*

(1) *Let* U *be the subgroup of* $\mathrm{Aut}(\mathrm{D}^b(\mathbb{X}))$ *which is generated by the translation* T *and all tubular mutations associated to exceptional tubes (for all slopes). Then* U *acts with three orbits on the set of all slope categories.*

(2) *Moreover, for* $\mathrm{K}_0(\mathbb{X})$ *there are three possible cases:*

a) *The symbol is* $\begin{pmatrix} 2 \\ 4 \\ 2 \end{pmatrix}$. *Then the index of* $\mathbb{X}$ *is one, two or three (the precise value depending on the arithmetic of* k).

b) *The symbol is* $\begin{pmatrix} 2 \\ 4 \\ 4 \end{pmatrix}$ *or* $\left(\begin{array}{c} 2 \\ 2 \end{array} \middle| 2 \right)$. *Then the index of* $\mathbb{X}$ *is two or three.*

c) *The symbol is* $\begin{pmatrix} 2 \\ 4 \end{pmatrix}$ *or* $\begin{pmatrix} 2 \\ 2 \\ 2 \end{pmatrix} 2$. *Then the index of* $\mathbb{X}$ *is two or three.* $\square$

REMARK 8.2.9. (1) For the case c) in the preceding proposition we constructed examples for indices two and three, respectively ([**59**]; see the following sections). We remark that in [**59**] we stated that in cases a) and b) examples of index three do not exist. This was based on an argument which turned out to be wrong. We now do not see any reason why examples of index three in these cases should not exist.

(2) Which index actually occurs depends on the question which of the 6 cosets of the subgroup $\langle t, \ell, s^4 \rangle$ in B_3 have representatives which are realizable by automorphisms of the derived category. If (besides the identity) none of these representatives is realizable the index is three. If additionally only s^2 is realizable then the index is two. If any other coset is realizable then the index is one (only possible in case a). Another question (related to Problem 5.2.2) is whether such realizations (assuming existence) are always possible by tubular mutations (compare Proposition 6.3.4).

8.3. A tubular curve of index three

In this section we exhibit our example [**59**] of a tubular exceptional curve $\mathbb{X}$ of index three and list its further properties. Knowledge of the action of $\mathrm{Aut}(\mathrm{D}^b(\mathbb{X}))$ on the rational helix allows to determine $\mathrm{Aut}(\mathrm{D}^b(\mathbb{X}))$ itself.

Recall that for sequences $\mathbf{p} = (p_1, \ldots, p_t)$ and $\mathbf{d} = (d_1, \ldots, d_t)$ of positive integers the abelian group $\mathbb{Z}\left[\frac{d_1}{p_1}, \ldots, \frac{d_t}{p_t}\right]$ is also denote by $\mathbb{L}(\mathbf{p}, \mathbf{d})$.

PROPOSITION 8.3.1. *There is a tubular exceptional curve* $\mathbb{X}$ *over* $k = \mathbb{Q}$ *such that the following holds*:

(1) *The index of* $\mathbb{X}$ *is three.*

(2) *A projective coordinate algebra of* $\mathbb{X}$ *is given by the graded factorial algebra*

$$R = \mathbb{Q}[X, Y, Z, U]/(X^2 + Y^2 + Z^2, U^2 - X^2 - 3Y^2),$$

which is graded by $\mathbb{L}(\mathbf{p}, \mathbf{d})$ *with* $\mathbf{p} = (1,1,1,2)$ *and* $\mathbf{d} = (1,1,1,2)$. (*This property uniquely determines* $\mathbb{X}$.)

(3) *There is a tilting bundle whose endomorphism ring is the canonical algebra* Λ *given by the species*

$$\begin{array}{ccc} & K & \\ & {}^{K}\nearrow \quad \searrow{}^{M} & \\ \mathbb{Q} \xrightarrow{\quad F \quad} & & F \end{array}$$

plus certain relations, where $K = \mathbb{Q}(\sqrt{-3}, \sqrt{2})$, $F = \left(\frac{-1,-1}{\mathbb{Q}}\right)$ *be the skew field of quaternions over* $\mathbb{Q}$ *on generators* $\mathbf{i}$ *and* $\mathbf{j}$ *with relations*

$$\mathbf{i}^2 = -1, \ \mathbf{j}^2 = -1, \ \mathbf{ij} = -\mathbf{ji},$$

and moreover, M *is the bimodule from 5.7.3, that is,* $M = {}_K(K \oplus K)_F$ *with the canonical* K-*action and the* F-*action defined by*

$$(x, y) \cdot \mathbf{i} = \frac{1}{\sqrt{-3}}(\sqrt{2}x + y, x - \sqrt{2}y), \ (x, y) \cdot \mathbf{j} = (y, -x)$$

for all $x, y \in K$.

(4) *For the automorphism groups we have* $\mathrm{Aut}(\mathbb{X}) \simeq \mathbb{V}_4$, *the Klein four group, and*

$$\mathrm{Aut}(\mathrm{D}^b(\mathbb{X})) \simeq (\mathbb{Z}_2 \times \mathbb{V}_4) \rtimes (F_2 \times \mathbb{Z}),$$

where F_2 is the free group in two generators.

PROOF. (2) Consider the tame bimodule ${}_kF_F$. The associated homogeneous exceptional curve admits $S = \mathbb{Q}[X,Y,Z]/(X^2 + Y^2 + Z^2) = \mathbb{Q}[x,y,z]$ by 5.5.1 as projective coordinate algebra. The element x^2+3y^2 is a prime element in S (see [**53**, 3.10.1]). Insertion of the weight $p = 2$ into this prime leads to the H-graded factorial algebra

$$R = \mathbb{Q}[X,Y,Z,U]/(X^2 + Y^2 + Z^2, U^2 - X^2 - 3Y^2),$$

where $H = \mathbb{L}(\mathbf{p},\mathbf{d})$ is the abelian group as above, that is, generated by the degrees $\deg X$, $\deg Y$, $\deg Z$, $\deg U$ with relations $\deg X = \deg Y = \deg Z$, $2\deg U = 2\deg X$. The torsion subgroup of H is generated by $\deg U - \deg X$. Denote by $\mathbb{X}$ the corresponding exceptional curve with hereditary category

$$\mathcal{H} = \mathrm{coh}(\mathbb{X}) = \frac{\mathrm{mod}^H(R)}{\mathrm{mod}_0^H(R)}.$$

Obviously, $\mathbb{X}$ has the symbol $\sigma[\mathbb{X}] = \begin{pmatrix} 2 \\ 2 \\ 2 \end{pmatrix} \begin{matrix} 2 \end{matrix}$, and hence is tubular.

(3) The construction in [**70**, Prop. 5.4] leads to a tilting bundle $L \longrightarrow L_1(1) \longrightarrow \overline{L}$ whose endomorphism ring is the canonical algebra Λ as described in (3): The endomorphism ring of the simple object, which corresponds to the prime element $x^2 + 3y^2$, is isomorphic to $\mathbb{Q}[X,Y]/(X^2+Y^2+1, X^2+3Y^2) \simeq \mathbb{Q}(\sqrt{-3},\sqrt{2})$. Moreover, by considering dimensions (see [**66**, Prop. 10.1]), M is a $(2,2)$-bimodule, moreover a simple bimodule, since $K \not\simeq F$. Considering the isomorphism of algebras,

$$K \otimes_\mathbb{Q} F \simeq \left(\frac{-1,-1}{K}\right) \simeq \mathrm{M}_2(K),$$

using that M is a simple $K \otimes_\mathbb{Q} F$-module, it follows, that M is the bimodule from 5.7.3.

(1) Since L, $L(1)$ and $\overline{L}$ are exceptional having pairwise non-isomorphic endomorphism skew fields $\mathbb{Q}$, K and F, respectively, the three corresponding tubular families are pairwise non-equivalent, hence the index of $\mathbb{X}$ is three. In fact, each tubular family contains precisely one exceptional tube (of rank two) and therefore contains precisely two exceptional objects, having the same endomorphism skew field.

(4) Since the ghost group $\mathcal{G}$ is trivial, from 8.1.10 we get the following exact sequence of groups

$$1 \longrightarrow \mathrm{Pic}_0(\mathbb{X}) \rtimes \mathrm{Aut}(\mathbb{X}) \longrightarrow \mathrm{Aut}(\mathrm{D}^b(\mathbb{X})) \overset{\Phi}{\longrightarrow} B_3,$$

where $\mathrm{Pic}_0(\mathbb{X})$ is isomorphic to the torsion part of the grading group H, hence to $\mathbb{Z}_2$. Moreover, $\mathrm{Aut}(\mathbb{X})$ is the automorphism group of $\mathbb{X}$, which by 6.3.1 consists of the automorphisms of the projective spectrum $\mathbb{Y}$ of the graded factorial algebra S fixing the point corresponding to the prime element $x^2 + 3y^2$. By 5.5.1, $\mathrm{Aut}(\mathbb{Y}) \simeq \mathrm{SO}_3(\mathbb{Q})$, and this group acts on prime elements of degree one (which are of the form $\alpha x + \beta y + \gamma z$) like a matrix on (α,β,γ). In particular, each automorphism of $\mathbb{Y}$ is uniquely determined by its action on the points of degree one. Each such automorphism thus extends uniquely to a graded algebra automorphism of S. Then it is easy to check, which automorphisms are fixing the prime ideal generated by $x^2 + 3y^2$, and this yields $\mathrm{Aut}(\mathbb{X}) \simeq \mathbb{V}_4$, the Klein four group.

We determine the image of Φ: Consider the tubular mutations σ_L and σ_S which are associated to the tube containing L (of slope 0) and to the tube containing an exceptional simple object S (of slope ∞), respectively. By [**57**, 6.] these maps on slopes induce the actions $q \mapsto q/(1-2q)$ and $q \mapsto q+2$, respectively. It follows that the image of Φ is given by $\langle t, s^2, \ell^2 \rangle$ (with $t = (s\ell)^3$), which is a subgroup in B_3 of index 6. It follows from [**100**] that $\langle s^2, \ell^2 \rangle$ is isomorphic to F_2 and t is central, hence we get $\operatorname{Im}\Phi \simeq F_2 \times \mathbb{Z}$. It is easy to see, that the induced exact sequence of groups

$$1 \longrightarrow \operatorname{Pic}_0(\mathbb{X}) \rtimes \operatorname{Aut}(\mathbb{X}) \longrightarrow \operatorname{Aut}(\mathrm{D}^b(\mathbb{X})) \xrightarrow{\Phi} \operatorname{Im}\Phi \longrightarrow 1$$

splits. Then the result follows. $\qquad\square$

8.3.2. We keep the notation of the proposition. There are two companion curves which are derived equivalent to $\mathbb{X}$, namely the curves $\mathbb{X}\langle 0 \rangle$ and $\mathbb{X}\langle 1 \rangle$ (in the slopes $q = 0$ and $q = 1$, respectively). Moreover, $\mathbb{X}$, $\mathbb{X}\langle 0 \rangle$ and $\mathbb{X}\langle 1 \rangle$ are pairwise non-isomorphic.

This corresponds to the fact, that there are two (further) tilting bundles in $\mathcal{H}$, such that the endomorphism rings are the canonical algebras $\Lambda\langle 0 \rangle$ and $\Lambda\langle 1 \rangle$ given by the species

$$\begin{array}{ccc} & \mathbb{Q} & \\ F^* \nearrow & & \searrow K \\ F \xrightarrow{\quad M^* \quad} & & K \end{array}$$

(where F^* and M^* denote the dual bimodules) and

$$\begin{array}{ccc} & F & \\ F \nearrow & & \searrow N \\ \mathbb{Q} \xrightarrow{\quad K \quad} & & K \end{array}$$

(plus relations), respectively (for some bimodule N), see [**59**]. The algebras Λ, $\Lambda\langle 0 \rangle$, and $\Lambda\langle 1 \rangle$ are tilting equivalent.

It follows from 1.7.12 that for $\mathbb{X}\langle 1 \rangle$ one gets as projective coordinate algebra a graded algebra, arising by inserting the weight $p = 2$ into some prime element of

$$\mathbb{Q}\langle X, Y, Z \rangle/(XY - YX, XZ - ZX, YZ + ZY, Z^2 - 3Y^2 - 2X^2).$$

8.3.3. In [**73**] the following is shown over an algebraically closed field: Two finite dimensional algebras Λ and Λ' which are derived equivalent to the same tubular exceptional curve and having the same Cartan matrix are isomorphic.

This is not true in general over arbitrary fields, since the tilting equivalent tubular canonical algebras Λ and $\Lambda\langle 1 \rangle$ as above have the same Cartan matrix, but are obviously not isomorphic.

8.4. A related tubular curve of index two

The next example shows that the index is not a K-theoretic invariant.

PROPOSITION 8.4.1. *There is a tubular exceptional curve $\mathbb{X}'$ over the field $k = \mathbb{Q}(\mathbf{i})$ such that the following holds:*

(1) *With the tubular curve $\mathbb{X}$ from 8.3.1, the Grothendieck groups $\mathrm{K}_0(\mathbb{X}')$ and $\mathrm{K}_0(\mathbb{X})$, equipped with the Euler form, are isomorphic.*

(2) *The index of $\mathbb{X}'$ is two.*

(3) *There is a graded factorial coordinate algebra of $\mathbb{X}'$ which arises by insertion of the weight $p = 2$ into the central prime element $X^4 - Y^4$ in the twisted polynomial algebra $K[X; Y, \alpha]$, where $K = k(\sqrt[4]{2})$ and α is the k-automorphism $\sqrt[4]{2} \mapsto \mathbf{i}\sqrt[4]{2}$.*

(4) *There is a tilting bundle whose endomorphism ring is the canonical algebra* Λ' *given by the species*

$$
\begin{array}{ccc}
 & K \nearrow \overset{k}{} \searrow K & \\
K & \xrightarrow{\quad M \quad} & K
\end{array}
$$

plus certain relations, where M is the non-simple bimodule $M(K, \alpha)$.

(5) $\mathrm{Aut}(\mathbb{X}') \simeq \mathbb{Z}_4$ *coincides with the ghost group (hence the geometric automorphism group is trivial), and for $\mathrm{Aut}(\mathrm{D}^b(\mathbb{X}'))$ there is the exact sequence of groups*

$$1 \longrightarrow \mathbb{Z}_2 \times \mathbb{Z}_4 \longrightarrow \mathrm{Aut}(\mathrm{D}^b(\mathbb{X}')) \longrightarrow U \longrightarrow 1,$$

where U is the subgroup $\langle s^2, \ell \rangle$ of the braid group B_3.

PROOF. (3) The field extension K/k is Galois with cyclic Galois group generated by α. Let $\mathbb{X}_0'$ be the homogeneous curve associated to the bimodule $M = M(K, \alpha)$. A projective coordinate algebra is given by $K[X; Y, \alpha]$. Insertion of the weight $p = 2$ into the central prime element $\pi = X^4 - Y^4$ leads by 6.2.4 to an exceptional curve $\mathbb{X}'$. By 1.7.10 the multiplicity of the inserted point is 4, and hence the symbol of $\mathbb{X}'$ is $\binom{2}{4}$ and $\mathbb{X}'$ is tubular.

(1) This follows, since the symbols $\binom{2}{4}$ and $\left(\begin{array}{c|c} 2 \\ 2 \\ 2 \end{array} \, 2 \right)$ are equivalent [**57**].

(2) Since M is non-simple, by 8.2.5 the index is two.

(4) This follows, since by 1.7.10 the endomorphism ring of the exceptional simple object is k.

(5) By 5.3.4, $\mathcal{G} \simeq \mathbb{Z}_4$, and for any transformation $Y \mapsto aY$ on $\mathbb{X}_0'$ with $a \in \mathbb{Q}(\mathbf{i})^*$ leaving the prime ideal generated by $X^4 - Y^4$ fixed, $N(a) = 1$ follows and $Y \mapsto aY$ is trivial. Thus $\mathrm{Aut}(\mathbb{X}') = \mathcal{G}$.

Consider the map $\Phi : \mathrm{Aut}(\mathrm{D}^b(\mathbb{X}')) \longrightarrow B_3$. Take the tubular mutations σ_L and σ_S which are associated to the tube containing L (of slope 0) and to a homogeneous tube (containing a simple object S of slope ∞), respectively. By [**57**, 6.] these maps on slopes induce the actions $q \mapsto q/(1-q)$ and $q \mapsto q + 2$, respectively. It follows that the image of Φ is given by $\langle s^2, \ell \rangle$. (This subgroup in B_3 is of index three, and described by the defining relation $(s^2 \ell)^2 = (\ell s^2)^2$, compare [**58**].)

By 8.1.10 the kernel of Φ is generated by $\mathcal{G}$ and the generator of $\mathrm{Pic}_0(\mathbb{X}')$ of order two. By 0.4.8 this generator commutes with all ghosts, and thus the kernel of Φ is given by $\mathrm{Pic}_0(\mathbb{X}') \times \mathcal{G}$. $\square$

8.5. Line bundles which are not exceptional

EXAMPLE 8.5.1. Let $k = \mathbb{R}$ and let $\mathbb{X}$ be the tubular exceptional curve arising by inserting the weight $p = 2$ into the central prime element $\pi = X^4 + Y^4$ in $\mathbb{C}[X, \overline{Y}]$. Then there is a line bundle L', which lies in a homogeneous tube.

PROOF. Let S be one (of the two) exceptional simple object in $\mathcal{H}$ such that $\mathrm{Hom}(L, S) \neq 0$. We get a short exact sequence

$$0 \longrightarrow L' \longrightarrow L \longrightarrow S \longrightarrow 0,$$

where L' is a line bundle. By 5.6.1 the exceptional point x has symbol data $f(x) = 2$, $e(x) = 2$, and then $\langle [L'], [L'] \rangle = 0$. But then $\mathrm{Ext}^1(L', L') \neq 0$. Since line bundles are stable, L' lies in the mouth of a homogeneous tube. $\square$

This fact is surprising, since in the commutative case graded factoriality implies that each line bundle is a degree shift of the structure sheaf L [**34**, Prop. 2.1]. We conclude, that this is not longer true in the noncommutative case, since the projective coordinate algebra R, which arises by inserting the weight $p = 2$ into $\pi = X^4 + Y^4$ in $\mathbb{C}[X, \overline{Y}]$ is graded factorial.

8.5.2. We construct further examples of non-exceptional line bundles.

(1) In the same way as in 8.5.1 such a line bundle can be constructed for any tubular exceptional curve with symbol $\left(\begin{smallmatrix} 2 \\ 2 \end{smallmatrix} \, \middle| \, 2 \right)$ or $\begin{pmatrix} 2 \\ 4 \\ 2 \end{pmatrix}$, and similarly (taking S^2 as cokernel), for the symbol $\begin{pmatrix} 2 \\ 4 \end{pmatrix}$.

(2) Consider a tubular curve $\mathbb{X}$ with symbol $\begin{pmatrix} 3 \\ 3 \end{pmatrix}$. (Compare also Example 1.6.10.) For the kernel L' of an epimorphism $L \longrightarrow S \oplus S^{(2)}$ one gets $\langle [L'], [L'] \rangle = 0$, where S is exceptional simple and $S^{(2)}$ is the indecomposable middle term of the almost split sequence ending in S. Such an epimorphism exists: The universal extension over the underlying homogeneous curve

$$0 \longrightarrow L(-x) \xrightarrow{x_L} L \longrightarrow S_x^3 \longrightarrow 0$$

and the projections $S_x^3 \twoheadrightarrow S_x^2 \twoheadrightarrow S_x$ lead to the epimorphism of 3-cycles

$$
j(L) = \Bigg[\quad
\begin{array}{ccccc}
L & \!\!=\!\! & L & \!\!=\!\! & L & \xrightarrow{x_L} & L(x) \\
\downarrow & & \downarrow & & \downarrow & & \downarrow \\
S_x^2 & \!\!\longrightarrow\!\!\! & S_x & \longrightarrow & 0 & \longrightarrow & S_x^2(x)
\end{array}
\quad \Bigg]
$$

$$S \oplus S^{(2)} = \Big[\qquad \qquad \qquad \qquad \qquad \qquad \Big]$$

Representing L' as kernel, we see that L' stores an "irreducible" factorization of x_L.

(3) For the symbol $\begin{pmatrix} 2 & 2 \\ 2 & 2 \end{pmatrix}$ it is also possible to construct an example. Take again $\mathbb{C}[X, \overline{Y}]$ and take the central prime elements $\pi_1 = Y^2 - X^2 = (Y - X)(Y + X)$ and $\pi_2 = Y^2 - 4X^2 = (Y - 2X)(Y + 2X)$. The element $(Y + X)(Y - 2X)$ induces a short exact sequence

$$0 \longrightarrow L(-2) \longrightarrow L \longrightarrow S_1 \oplus S_2 \longrightarrow 0,$$

where S_1 and S_2 are the simple objects concentrated in the points x_1 and x_2 associated to π_1 and π_2. This induces, after insertion of weights, a short exact sequence in $\mathcal{H} \begin{pmatrix} 2 & 2 \\ x_1 & x_2 \end{pmatrix}$

$$0 \longrightarrow L' \longrightarrow j(L) \longrightarrow S_1' \oplus S_2' \longrightarrow 0,$$

where S_1' and S_2' are exceptional simple objects concentrated in different points and L' is a line bundle. It is easy to see, that $\langle [L'], [L'] \rangle = 0$.

PROPOSITION 8.5.3. *Let $\mathbb{X}$ be a tubular exceptional curve. If there is an exceptional point which is multiplicity free then each line bundle is exceptional. Accordingly, if there exists a line bundle L' which is not exceptional then the symbol of $\mathbb{X}$ is one of the five symbols in 8.5.2.*

PROOF. For an exceptional point x call a line bundle L' x-*special*, if L' maps onto precisely one simple object S_x in $\mathcal{U}_x$. If $e(x) = 1$, then each line bundle L' is x-special, which follows from the formula

$$1 = \mathrm{rk}(L') = \frac{1}{\kappa\varepsilon f(x)}\left(\langle[L'],[S_x]\rangle + \cdots + \langle[L'],[\tau^{p(x)-1}S_x]\rangle\right),$$

and each summand inside the brackets is divisible by $\langle[S_x],[S_x]\rangle = \kappa\varepsilon f(x)$ (compare 0.4.5).

If L' is x-special for some exceptional point x, then L' is exceptional, since otherwise it would lie in a homogeneous tube and would be therefore τ-stable. Therefore we can exclude all tubular curves where there is a multiplicity free, exceptional point, which one can detect from the symbol. Only for the five symbols from 8.5.2 such a point does not exist.   □

REMARK 8.5.4. Let L' be a line bundle. If L is special with $\deg(L) > \deg(L')$, then by the Riemann-Roch formula L' embeds into L (up to τ-translations). If $\deg(L)$ is minimal with this property then it is easy to check K-theoretically whether $[L']$ is a root or not. One gets that the examples above are essentially *all* examples of non-exceptional line bundles (in the tubular case).

REMARK 8.5.5. Let $\mathbb{X}$ be the homogeneous exceptional curve with projective coordinate algebra $R = \mathbb{C}[X, \overline{Y}]$. Consider the non-exceptional line bundle L' as 2-cycle, concentrated in the point x corresponding to $\pi = X^4 + Y^4$, which decomposes as $\pi = (X^2 - \mathbf{i}Y^2)(X^2 + \mathbf{i}Y^2)$ into irreducible elements u_1 and u_2:

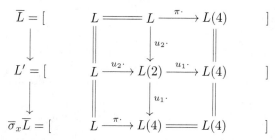

This construction can be done generally with an irreducible decomposition of a central prime (see 6.2.7). It would be interesting to understand, why this leads sometimes to exceptional line bundles L' and sometimes to non-exceptional L'.

Automorphism groups over the real numbers

A.1. Tables for the domestic and tubular cases

If k is algebraically closed and of tubular weight type (2 2 2 2), then $\mathbb{X}$ depends also on some parameter $\lambda \in k$, $\lambda \neq 0, 1$. More precisely, two such curves $\mathbb{X}(2\,2\,2\,2; \lambda)$ and $\mathbb{X}(2\ 2\ 2\ 2; \mu)$ are isomorphic if and only if they have the same j-invariant $j(\lambda) = 2^8(\lambda^2 - \lambda + 1)^3/(\lambda^2(\lambda - 1)^2)$. Moreover, the automorphism group depends on this j-invariant [72]:

$$\operatorname{Aut}\mathbb{X} = \begin{cases} \mathbb{A}_4 & j = 0, \\ \mathbb{D}_4 & j = 1728, \\ \mathbb{V}_4 & j \neq 0, 1728. \end{cases}$$

Here, $\mathbb{A}_4$ denotes the alternating group (which is of order 12), $\mathbb{D}_4$ the dihedral group (of order 8) and $\mathbb{V}_4 = \mathbb{Z}_2 \times \mathbb{Z}_2$ the Klein four group. In Table A.2 we denote by $\nabla \operatorname{PGL}_2(\mathbb{R})$ the subgroup of $\operatorname{PGL}_2(\mathbb{R})$ formed by the upper triangular matrices.

In the following tables (taken from [58]), we exhibit the automorphism groups for the domestic and for the tubular exceptional curves over $\mathbb{R}$. When parameters t occur, then a fundamental domain of these parameters is indicated. (For the determination of these parameter domains we refer to [58].) M denotes the underlying tame bimodule. The pictures in the tables indicate which weight is inserted in which type of point(s) on the quotient of the Riemann sphere (with boundary). By 6.3.1 one has to check which automorphisms fix the given point(s). For the "classical" case, where $M = \mathbb{C} \oplus \mathbb{C}$, we refer to [72]; it is not treated again.

The letters a, b, c, d, e in the tables indicate derived equivalence: Each tubular curve in the table with a letter from a, b, c, d, e is derived equivalent to a curve in another table with the same letter.

In case $M = \mathbb{C} \oplus \overline{\mathbb{C}}$ it is easy to determine the geometric automorphism group $\operatorname{Aut}(\mathbb{X})/\mathcal{G}$ from $\operatorname{Aut}(\mathbb{X})$; the only ghost automorphism is induced by complex conjugation which is of order two.

From the tables we deduce the following

PROPOSITION A.1.1. (1) *There are no parameters in the domestic cases.*
(2) *If $\mathbb{X}$ is tubular then* $\operatorname{Aut}\mathbb{X}$ *is finite.*

Case	Symbol	Parameter	Aut $\mathbb{X}_t$
D 1	$(p \mid 2)$	$-$	$\mathbb{R}/2\pi \rtimes \mathbb{Z}_2$
T 1 c	$(2\ 2 \mid 2)$	$t \in [0,\ 1)$	$\mathbb{D}_4 \quad t=0$ $\mathbb{V}_4 \quad t \neq 0$

TABLE A.1. Domestic and tubular curves with $M = {}_{\mathbb{R}}\mathbb{H}_{\mathbb{H}}$

Case	Weights	Symbol	$\mathbb{R} \oplus \mathbb{R}$	$\mathbb{H} \oplus \mathbb{H}$
D 1	p	(p)	$\nabla\, \mathrm{PGL}_2(\mathbb{R})$	$\nabla\, \mathrm{PGL}_2(\mathbb{R})$
D 2	p	$\left(\begin{smallmatrix} p \\ 2 \\ p \\ 2 \end{smallmatrix}\right)$	$-$ $\mathbb{C}^*/\mathbb{R}^* \rtimes \mathbb{Z}_2$	$\mathbb{C}^*/\mathbb{R}^* \rtimes \mathbb{Z}_2$ $-$
D 3	$p_1\ p_2$	$(p_1\ p_2)$	$\mathbb{R}^* \rtimes \mathbb{Z}_2 \quad p_1 = p_2$ $\mathbb{R}^* \quad\quad p_1 \neq p_2$	$\mathbb{R}^* \rtimes \mathbb{Z}_2 \quad p_1 = p_2$ $\mathbb{R}^* \quad\quad p_1 \neq p_2$
D 4	n 2	$\left(\begin{smallmatrix} 2 & n \\ 2 & 1 \\ 2 & n \\ 2 & 1 \end{smallmatrix}\right)$	$-$ $\mathbb{Z}_2$	$\mathbb{Z}_2$ $-$
D 5	2 3	$\left(\begin{smallmatrix} 2 & 3 \\ 1 & 2 \\ 2 & 3 \\ 1 & 2 \end{smallmatrix}\right)$	$-$ $\mathbb{Z}_2$	$\mathbb{Z}_2$ $-$
D 6	2 n 2	$(2\ 2\ n)$	$\mathbb{Z}_2 \quad n > 2$ $\mathbb{S}_3 \quad n = 2$	$\mathbb{Z}_2 \quad n > 2$ $\mathbb{S}_3 \quad n = 2$
D 7	3 3 2	$(2\ 3\ 3)$	$\mathbb{Z}_2$	$\mathbb{Z}_2$
D 8	3 4 2	$(2\ 3\ 4)$	1	1
D 9	3 5 2	$(2\ 3\ 5)$	1	1

TABLE A.2. Domestic curves with $M = \mathbb{R} \oplus \mathbb{R}$ and $M = \mathbb{H} \oplus \mathbb{H}$

Case	Weights		Symbol	Aut $\mathbb{X}$
	$\mathbb{H}$	$\mathbb{R}$		
D 1			(p)	$\mathbb{R}_+ \times \mathbb{Z}_2$
D 2			$\left(\begin{smallmatrix} p \\ 2 \end{smallmatrix}\right)$	$\mathbb{V}_4$
D 3			$\left(\begin{smallmatrix} p \\ 2 \end{smallmatrix}\right)$	$\mathbb{V}_4$
D 4			$(p_1 \; p_2)$	$\mathbb{R}_+ \rtimes \mathbb{V}_4 \quad p_1 = p_2$ $\mathbb{R}_+ \times \mathbb{Z}_2 \quad p_1 \neq p_2$
D 5			$\left(\begin{smallmatrix} 2 & n \\ 2 & 1 \end{smallmatrix}\right)$	$\mathbb{Z}_2$
D 6			$\left(\begin{smallmatrix} 2 & n \\ 2 & 1 \\ 2 & 1 \end{smallmatrix}\right)$	$\mathbb{Z}_2$
D 7			$\left(\begin{smallmatrix} 2 & 3 \\ 1 & 2 \end{smallmatrix}\right)$	$\mathbb{Z}_2$
D 8			$\left(\begin{smallmatrix} 2 & 3 \\ 1 & 2 \\ 1 & 2 \end{smallmatrix}\right)$	$\mathbb{Z}_2$

TABLE A.3. Domestic curves with $M = \mathbb{C} \oplus \overline{\mathbb{C}}$

Case	Weights	Symbol	$\mathbb{R}\oplus\mathbb{R}$	$\mathbb{H}\oplus\mathbb{H}$
b T 1 a		$\left(\begin{smallmatrix}2&2\\2&2\end{smallmatrix}\right)$ $\left(\begin{smallmatrix}2&2\\2&2\end{smallmatrix}\right)$	— $\mathbb{V}_4\ \ t\in(0,1)$	$\mathbb{V}_4\ \ t\in(0,1)$ —
e T 2 d		$\left(\begin{smallmatrix}2&4\\1&2\end{smallmatrix}\right)$ $\left(\begin{smallmatrix}2&4\\1&2\end{smallmatrix}\right)$	— $\mathbb{Z}_2$	$\mathbb{Z}_2$ —
T 3		$\left(\begin{smallmatrix}3&3\\1&2\end{smallmatrix}\right)$ $\left(\begin{smallmatrix}3&3\\1&2\end{smallmatrix}\right)$	— $\mathbb{Z}_2$	$\mathbb{Z}_2$ —
T 4		$(2\ 3\ 6)$	1	1
T 5		$(2\ 4\ 4)$	$\mathbb{Z}_2$	$\mathbb{Z}_2$
T 6		$(3\ 3\ 3)$	$\mathbb{S}_3$	$\mathbb{S}_3$
T 7		$\left(\begin{smallmatrix}2&2&2\\1&1&2\end{smallmatrix}\right)$ $\left(\begin{smallmatrix}2&2&2\\1&1&2\end{smallmatrix}\right)$	— $\mathbb{Z}_2\ \ t=\pi/2;$ $1\quad t\in(0,\pi)$ $\quad t\neq\pi/2$	$\mathbb{Z}_2\ \ t=\pi/2;$ $1\quad t\in(0,\pi)$ $\quad t\neq\pi/2$ —
T 8		$(2\ 2\ 2\ 2)$	$\mathbb{A}_4\ \ j=0$ $\mathbb{D}_4\ \ j=1728$ $\mathbb{V}_4\ \ j\neq0,1728$	$\mathbb{A}_4\ \ j=0$ $\mathbb{D}_4\ \ j=1728$ $\mathbb{V}_4\ \ j\neq0,1728$

TABLE A.4. Tubular curves for $M=\mathbb{R}\oplus\mathbb{R}$ and $M=\mathbb{H}\oplus\mathbb{H}$

Case	Weights		Symbol	Parameter	Aut $\mathbb{X}_t$
	$\mathbb{H}$	$\mathbb{R}$			
T 1			$\begin{pmatrix} 2 \\ 4 \\ 2 \end{pmatrix}$	$t \in (0, \pi)$	$\mathbb{V}_4$
T 2 a			$\begin{pmatrix} 2 & 2 \\ 2 & 2 \end{pmatrix}$	$t \in (0, 1)$	$\mathbb{V}_4$
T 3 b			$\begin{pmatrix} 2 & 2 \\ 2 & 2 \\ 2 & 2 \end{pmatrix}$	$t \in (0, 1)$	$\mathbb{V}_4$
T 4 c			$\begin{pmatrix} 2 & 2 \\ 2 & 2 \\ 1 & 2 \end{pmatrix}$	$t \in (0, 1]$	$\mathbb{V}_4 \quad t = 1$ $\mathbb{Z}_2 \quad t \neq 1$
T 5 d			$\begin{pmatrix} 2 & 4 \\ 1 & 2 \end{pmatrix}$	$-$	$\mathbb{Z}_2$
T 6 e			$\begin{pmatrix} 2 & 4 \\ 1 & 2 \\ 1 & 2 \end{pmatrix}$	$-$	$\mathbb{Z}_2$
T 7			$\begin{pmatrix} 3 & 3 \\ 1 & 2 \end{pmatrix}$	$-$	$\mathbb{Z}_2$
T 8			$\begin{pmatrix} 3 & 3 \\ 1 & 2 \\ 1 & 2 \end{pmatrix}$	$-$	$\mathbb{Z}_2$
T 9			$\begin{pmatrix} 2 & 2 & 2 \\ 1 & 1 & 2 \end{pmatrix}$	$-$	$\mathbb{V}_4$
T 10			$\begin{pmatrix} 2 & 2 & 2 \\ 1 & 1 & 2 \\ 1 & 1 & 2 \end{pmatrix}$	$-$	$\mathbb{V}_4$

TABLE A.5. Tubular curves with $M = \mathbb{C} \oplus \overline{\mathbb{C}}$

The tubular symbols

$$\left(\begin{smallmatrix}2\\4\end{smallmatrix}\right), \left(\begin{smallmatrix}2\\2\\2\end{smallmatrix}\;\middle|\;2\right) \qquad\qquad \left(\begin{smallmatrix}2&4\\1&2\end{smallmatrix}\right), \left(\begin{smallmatrix}2&4\\1&2\\1&2\end{smallmatrix}\right)$$

$$\left(\begin{smallmatrix}2\\4\\2\end{smallmatrix}\right) \qquad\qquad \left(\begin{smallmatrix}3&3\\1&2\end{smallmatrix}\right)$$

$$\left(\begin{smallmatrix}2\\4\\4\end{smallmatrix}\right), \left(\begin{smallmatrix}2\\2\end{smallmatrix}\;\middle|\;2\right) \qquad\qquad \left(\begin{smallmatrix}3&3\\1&2\\1&2\end{smallmatrix}\right)$$

$$\left(\begin{smallmatrix}3\\3\end{smallmatrix}\right), \left(\begin{smallmatrix}3\\3\\3\end{smallmatrix}\right) \qquad\qquad (2\ 3\ 6)$$

$$\left(\begin{smallmatrix}2&2\\1&3\end{smallmatrix}\right) \qquad\qquad (2\ 4\ 4)$$

$$\left(\begin{smallmatrix}2&2\\1&3\\1&3\end{smallmatrix}\right) \qquad\qquad (3\ 3\ 3)$$

$$\left(\begin{smallmatrix}2&2\\2&2\end{smallmatrix}\right), \left(\begin{smallmatrix}2&2\\2&2\\2&2\end{smallmatrix}\right) \qquad\qquad \left(\begin{smallmatrix}2&2&2\\1&1&2\end{smallmatrix}\right)$$

$$\left(\begin{smallmatrix}2&2\\2&2\\1&2\end{smallmatrix}\right), (2\,2\mid 2) \qquad\qquad \left(\begin{smallmatrix}2&2&2\\1&1&2\\1&1&2\end{smallmatrix}\right)$$

$$(2\ 2\ 2\ 2)$$

TABLE B.1. The 17 equivalence classes of tubular symbols

See 0.4.5 for the definition of symbols. Two symbols are called equivalent if they yield the same Grothendieck group with Euler form. The 17 boxes show the 17 equivalence classes. We refer to [57] for details.

Bibliography

1. S. A. Amitsur, *Prime rings having polynomial identities with arbitrary coefficients*, Proc. London Math. Soc. (3) **17** (1967), 470–486. MR 36 #209
2. M. Artin and J. J. Zhang, *Noncommutative projective schemes*, Adv. Math. **109** (1994), no. 2, 228–287. MR 96a:14004
3. I. Assem, D. Simson, and A. Skowroński, *Elements of the representation theory of associative algebras. Vol. 1*, London Mathematical Society Student Texts, vol. 65, Cambridge University Press, Cambridge, 2006, Techniques of representation theory. MR 2006j:16020
4. M. F. Atiyah, *Vector bundles over an elliptic curve*, Proc. London Math. Soc. (3) **7** (1957), 414–452. MR 24 #A1274
5. M. Auslander, I. Reiten, and S. O. Smalø, *Representation theory of Artin algebras*, Cambridge Studies in Advanced Mathematics, vol. 36, Cambridge University Press, Cambridge, 1997, Corrected reprint of the 1995 original. MR 98e:16011
6. D. Baer, *Einige homologische Aspekte der Darstellungstheorie Artinscher Algebren*, Dissertation, Universität Paderborn, 1983.
7. D. Baer, W. Geigle, and H. Lenzing, *The preprojective algebra of a tame hereditary artin algebra*, Comm. Algebra **15** (1987), 425–457. MR 88i:16036
8. H. Bass, *Algebraic K-theory*, W. A. Benjamin, Inc., New York-Amsterdam, 1968. MR 40 #2736
9. A. Bondal and D. Orlov, *Reconstruction of a variety from the derived category and groups of autoequivalences*, Compositio Math. **125** (2001), no. 3, 327–344. MR 2001m:18014
10. A. I. Bondal, *Representations of associative algebras and coherent sheaves*, Izv. Akad. Nauk SSSR Ser. Mat. **53** (1989), no. 1, 25–44. MR 90i:14017
11. A. Braun and C. R. Hajarnavis, *Finitely generated P.I. rings of global dimension two*, J. Algebra **169** (1994), no. 2, 587–604. MR 95j:16014
12. A. W. Chatters, *Noncommutative unique factorization domains*, Math. Proc. Cambridge Philos. Soc. **95** (1984), no. 1, 49–54. MR 85b:16001
13. A. W. Chatters and D. A. Jordan, *Noncommutative unique factorisation rings*, J. London Math. Soc. (2) **33** (1986), no. 1, 22–32. MR 87e:16001
14. P. M. Cohn, *Noncommutative unique factorization domains*, Trans. Amer. Math. Soc. **109** (1963), 313–331. MR 27 #5785
15. ———, *A remark on matrix rings over free ideal rings*, Proc. Cambridge Philos. Soc. **62** (1966), 1–4. MR 32 #5694
16. ———, *Free rings and their relations*, second ed., London Mathematical Society Monographs, vol. 19, Academic Press Inc. [Harcourt Brace Jovanovich Publishers], London, 1985. MR 87e:16006
17. W. W. Crawley-Boevey, *On tame algebras and bocses*, Proc. London Math. Soc. (3) **56** (1988), no. 3, 451–483. MR 89c:16028
18. ———, *Regular modules for tame hereditary algebras*, Proc. London Math. Soc. (3) **62** (1991), no. 3, 490–508. MR 92b:16024
19. ———, *Tame algebras and generic modules*, Proc. London Math. Soc. (3) **63** (1991), no. 2, 241–265. MR 92m:16019
20. ———, *Exceptional sequences of representations of quivers*, Proceedings of the Sixth International Conference on Representations of Algebras (Ottawa, ON, 1992) (Ottawa, ON), Carleton-Ottawa Math. Lecture Note Ser., vol. 14, Carleton Univ., 1992, p. 7. MR 94c:16017
21. ———, *Private email to the author*, February 2005.

122   BIBLIOGRAPHY

22. V. Dlab, *An introduction to diagrammatical methods in representation theory*, Vorlesungen aus dem Fachbereich Mathematik der Universität Essen, vol. 7, Universität Essen Fachbereich Mathematik, 1981, Lecture notes written by Richard Dipper. MR 83d:16030

23. _____, *The regular representations of the tame hereditary algebras*, Séminaire d`Algèbre Paul Dubreil et Marie-Paul Malliavin, Proceedings, Paris 1982 (35ème Année) (Berlin-Heidelberg-New York) (M.-P. Malliavin, ed.), Lecture Notes in Math., vol. 1029, Springer-Verlag, 1983, pp. 120–133. MR 85j:16040

24. V. Dlab and C. M. Ringel, *Indecomposable representations of graphs and algebras*, Mem. Amer. Math. Soc. **6** (1976), no. 173, v+57. MR 56 #5657

25. _____, *Normal forms of real matrices with respect to complex similarity*, Linear Algebra and Appl. **17** (1977), no. 2, 107–124. MR 57 #12552

26. _____, *Real subspaces of a quaternion vector space*, Canad. J. Math. **30** (1978), no. 6, 1228–1242. MR 80a:15033

27. _____, *The representations of tame hereditary algebras*, Representation Theory of Algebras. Proceedings of the Philadelphia Conference 1976 (New York) (R. Gordon, ed.), Marcel Dekker, 1978, pp. 329–353. Lecture Notes in Pure Appl. Math., Vol. 37. MR 58 #11021

28. _____, *The preprojective algebra of a modulated graph*, Representation Theory II (Proc. Second Internat. Conf. Carleton Univ., Ottawa, Ont., 1979) (Berlin-Heidelberg-New York), Lecture Notes in Math., vol. 832, Springer-Verlag, 1980, pp. 216–231. MR 83c:16022

29. _____, *A class of bounded hereditary noetherian domains*, J. Algebra **92** (1985), 311–321. MR 86h:16021

30. P. K. Draxl, *Skew fields*, London Mathematical Society Lecture Note Series, vol. 81, Cambridge University Press, Cambridge, 1983. MR 85a:16022

31. J.-M. Drezet and J. Le Potier, *Fibrés stables et fibrés exceptionnels sur* $\mathbf{P}_2$, Ann. Sci. École Norm. Sup. (4) **18** (1985), no. 2, 193–243. MR 87e:14014

32. Ju. A. Drozd, *Tame and wild matrix problems*, Representations and quadratic forms (Russian), Akad. Nauk Ukrain. SSR Inst. Mat., Kiev, 1979, pp. 39–74, 154. MR 82m:16028

33. R. M. Fossum, *The divisor class group of a Krull domain*, Ergebnisse der Mathematik und ihrer Grenzgebiete, vol. 74, Springer-Verlag, Berlin-Heidelberg-New York, 1973. MR 52 #3139

34. W. Geigle and H. Lenzing, *A class of weighted projective curves arising in representation theory of finite dimensional algebras*, Singularities, Representation of Algebras and Vector Bundles (Lambrecht 1985), Lecture Notes in Math., vol. 1273, Springer-Verlag, Berlin-Heidelberg-New York, 1987, pp. 265–297. MR 89b:14049

35. _____, *Perpendicular categories with applications to representations and sheaves*, J. Algebra **144** (1991), 273–343. MR 93b:16011

36. K. R. Goodearl and R. B. Warfield, *An introduction to noncommutative noetherian rings*, London Mathematical Society Student Texts, vol. 16, Cambridge University Press, Cambridge, 1989. MR 91c:16001

37. D. Happel, *Triangulated categories in the representation theory of finite dimensional algebras*, London Mathematical Society Lecture Note Series, no. 119, Cambridge University Press, Cambridge, 1988. MR 89e:16035

38. _____, *A characterization of hereditary categories with tilting object*, Invent. Math. **144** (2001), no. 2, 381–398. MR 2002a:18014

39. D. Happel and I. Reiten, *Hereditary abelian categories with tilting object over arbitrary base fields*, J. Algebra **256** (2002), no. 2, 414–432. MR 2004b:16016

40. D. Happel, I. Reiten, and S. Smalø, *Tilting in abelian categories and quasitilted algebras*, Mem. Amer. Math. Soc. **120** (1996), no. 575, viii+ 88. MR 97j:16009

41. D. Happel and C. M. Ringel, *Tilted algebras*, Trans. Amer. Math. Soc. **274** (1982), 399–443. MR 84d:16027

42. _____, *The derived category of a tubular algebra*, Representation Theory I. Finite Dimensional Algebras (Ottawa, Ont., 1984), Lecture Notes in Math., vol. 1177, Springer-Verlag, Berlin-Heidelberg-New York, 1986, pp. 156–180. MR 87j:18015

43. T. Hübner, *Exzeptionelle Vektorbündel und Reflektionen an Kippgarben über projektiven gewichteten Kurven*, Dissertation, Universität Paderborn, 1996.

44. N. Jacobson, *Finite-dimensional division algebras over fields*, Springer-Verlag, Berlin-Heidelberg-New York, 1996. MR 98a:16024

45. C. U. Jensen and H. Lenzing, *Model-theoretic algebra with particular emphasis on fields, rings, modules*, Algebra, Logic and Applications, vol. 2, Gordon and Breach Science Publishers, New York, 1989. MR 91m:03038

46. G. A. Jones and D. Singerman, *Complex functions. An algebraic and geometric viewpoint*, Cambridge University Press, Cambridge, 1987. MR 89b:30001

47. D. A. Jordan, *Unique factorisation of normal elements in noncommutative rings*, Glasgow Math. J. **31** (1989), no. 1, 103–113. MR 90e:16002

48. B. Keller, *On triangulated orbit categories*, Doc. Math. **10** (2005), 551–581 (electronic). MR 2007c:18006

49. O. Kerner, *Minimal approximations, orbital elementary modules, and orbit algebras of regular modules*, J. Algebra **217** (1999), no. 2, 528–554. MR 2000e:16018

50. H. Krause, *Generic modules over Artin algebras*, Proc. London Math. Soc. (3) **76** (1998), no. 2, 276–306. MR 98m:16017

51. _____, *The spectrum of a module category*, Mem. Amer. Math. Soc. **149** (2001), no. 707, x+125. MR 2001k:16010

52. S. A. Kuleshov and D. O. Orlov, *Exceptional sheaves on Del Pezzo surfaces*, Izv. Ross. Akad. Nauk Ser. Mat. **58** (1994), no. 3, 53–87. MR 95g:14048

53. D. Kussin, *Graduierte Faktorialität und die Parameterkurven tubularer Familien*, Dissertation, Universität Paderborn, 1997.

54. _____, *Factorial algebras, quaternions and preprojective algebras*, Algebras and Modules II (Geiranger, 1996) (I. Reiten, S. O. Smalø, and Ø. Solberg, eds.), CMS Conf. Proc., vol. 24, Amer. Math. Soc., Providence, RI, 1998, pp. 393–402. MR 99m:16020

55. _____, *Graded factorial algebras of dimension two*, Bull. London Math. Soc. **30** (1998), 123–128. MR 99d:13025

56. _____, *Non-isomorphic derived-equivalent tubular curves and their associated tubular algebras*, J. Algebra **226** (2000), 436–450. MR 2001d:16025

57. _____, *On the K-theory of tubular algebras*, Colloq. Math. **86** (2000), 137–152. MR 2001i:16014

58. _____, *The automorphism groups of domestic and tubular exceptional curves over the real numbers*, Representations of algebras. Vol. II (D. Happel and Y. B. Zhang, eds.), Beijing Norm. Univ. Press, Beijing, 2002, pp. 292–307. MR 2005f:16023

59. _____, *A tubular algebra with three types of separating tubular families*, Representations of algebras and related topics, Fields Inst. Commun., vol. 45, Amer. Math. Soc., Providence, RI, 2005, pp. 215–228. MR 2006k:16026

60. D. Kussin and H. Meltzer, *The braid group action for exceptional curves*, Arch. Math. (Basel) **79** (2002), 335–344. MR 2004d:16022

61. T. Y. Lam, *The algebraic theory of quadratic forms*, W. A. Benjamin, Inc., Reading, Mass., 1973, Mathematics Lecture Note Series. MR 53 #277

62. _____, *A first course in noncommutative rings*, Graduate Texts in Mathematics, vol. 131, Springer-Verlag, New York, 1991. MR 92f:16001

63. S. Lang, *Algebra*, third ed., Graduate Texts in Mathematics, Springer-Verlag, New York, 2002. MR 2003e:00003

64. H. Lenzing, *Curve singularities arising from the representation theory of tame hereditary artin algebras*, Representation Theory I. Finite Dimensional Algebras (Ottawa, Ont., 1984), Lecture Notes in Math., vol. 1177, Springer-Verlag, Berlin-Heidelberg-New York, 1986, pp. 199–231. MR 87i:16060

65. _____, *Wild canonical algebras and rings of automorphic forms*, Finite-dimensional algebras and related topics (Ottawa, ON, 1992), NATO Adv. Sci. Inst. Ser. C Math. Phys. Sci., vol. 424, Kluwer Acad. Publ., Dordrecht, 1994, pp. 191–212. MR 95m:16008

66. _____, *A K-theoretic study of canonical algebras*, Representation Theory of Algebras (Cocoyoc, 1994) (Providence, RI) (R. Bautista, R. Martínez-Villa, and J. A. de la Peña, eds.), CMS Conf. Proc., vol. 18, Amer. Math. Soc., 1996, pp. 433–473. MR 97e:16020

67. _____, *Hereditary noetherian categories with a tilting complex*, Proc. Amer. Math. Soc. **125** (1997), no. 7, 1893–1901. MR 98c:16013

68. _____, *Representations of finite dimensional algebras and singularity theory*, Trends in ring theory (Miskolc, 1996) (V. Dlab et al., ed.), CMS Conf. Proc., vol. 22, Amer. Math. Soc., Providence, RI, 1998, pp. 71–97. MR 99d:16014

69. _____, *Twenty-one characterizations of genus zero*, Mathematics & mathematics education (Bethlehem, 2000), World Sci. Publishing, River Edge, NJ, 2002, pp. 145–166. MR 2003h:14029

70. H. Lenzing and J. A. de la Peña, *Concealed-canonical algebras and separating tubular families*, Proc. London Math. Soc. (3) **78** (1999), no. 3, 513–540. MR 2000c:16018

71. H. Lenzing and H. Meltzer, *Sheaves on a weighted projective line of genus one, and representations of a tubular algebra*, Representations of Algebras (Ottawa, ON, 1992) (V. Dlab and H. Lenzing, eds.), CMS Conf. Proc., vol. 14, Amer. Math. Soc., Providence, RI, 1993, pp. 313–337. MR 94d:16019

72. _____, *The automorphism group of the derived category for a weighted projective line*, Comm. Algebra **28** (2000), no. 4, 1685–1700. MR 2001a:16021

73. _____, *Exceptional sequences determined by their Cartan matrix*, Algebr. Represent. Theory **5** (2002), no. 2, 201–209. MR 2003g:16012

74. H. Lenzing and I. Reiten, *Hereditary Noetherian categories of positive Euler characteristic*, Math. Z. **254** (2006), no. 1, 133–171. MR MR2232010

75. H. Lenzing and R. Zuazua, *Auslander-Reiten duality for abelian categories*, Bol. Soc. Mat. Mexicana (3) **10** (2004), no. 2, 169–177 (2005). MR 2006b:16025

76. H. Matsumura, *Commutative ring theory*, Cambridge Studies in Advanced Mathematics, vol. 8, Cambridge University Press, Cambridge, 1986. MR 88h:13001

77. J. C. McConnel and J. C. Robson, *Noncommutative Noetherian rings*, Pure and Applied Mathematics, John Wiley & Sons Ltd., Chichester, 1987. MR 89j:16023

78. H. Meltzer, *Exceptional sequences for canonical algebras*, Arch. Math. (Basel) **64** (1995), no. 4, 304–312. MR 96c:16022

79. _____, *Tubular mutations*, Colloq. Math. **74** (1997), no. 2, 267–274. MR 99a:18004

80. _____, *Exceptional vector bundles, tilting sheaves and tilting complexes for weighted projective lines*, Mem. Amer. Math. Soc. **171** (2004), no. 808, viii+139. MR 2005k:14033

81. B. Mitchell, *Theory of categories*, Pure and Applied Mathematics, Vol. XVII, Academic Press, New York, 1965. MR 34 #2647

82. J. Miyachi and A. Yekutieli, *Derived Picard groups of finite-dimensional hereditary algebras*, Compositio Math. **129** (2001), no. 3, 341–368. MR 2003c:18013

83. S. Mori, *Graded factorial domains*, Japan. J. Math. (N.S.) **3** (1977), no. 2, 223–238. MR 58 #27949

84. C. Năstăsescu and F. van Oystaeyen, *Graded ring theory*, North-Holland Mathematical Library, vol. 28, North-Holland Publishing Co., Amsterdam, 1982. MR 84i:16002

85. N. Popescu, *Abelian categories with applications to rings and modules*, Academic Press, London, 1973, London Mathematical Society Monographs, No. 3. MR 49 #5130

86. M. Prest, *Ziegler spectra of tame hereditary algebras*, J. Algebra **207** (1998), no. 1, 146–164. MR 2000c:16017

87. I. Reiten and M. Van den Bergh, *Grothendieck groups and tilting objects*, Algebr. Represent. Theory **4** (2001), no. 1, 1–23. MR 2002c:18007

88. _____, *Noetherian hereditary abelian categories satisfying Serre duality*, J. Amer. Math. Soc. **15** (2002), no. 2, 295–366 (electronic). MR 2003a:18011

89. C. M. Ringel, *Representations of K-species and bimodules*, J. Algebra **41** (1976), no. 2, 269–302. MR 54 #10340

90. _____, *Infinite dimensional representations of finite dimensional hereditary algebras*, Symposia Mathematica, Vol. XXIII (Conf. Abelian Groups and their Relationship to the Theory of Modules, INDAM, Rome, 1977), vol. 23, Academic Press, London, 1979, pp. 321–412. MR 81i:16032

91. _____, *Tame algebras and integral quadratic forms*, Lecture Notes in Mathematics, vol. 1099, Springer-Verlag, Berlin, 1984. MR 87f:16027

92. _____, *The canonical algebras*, Topics in Algebra, Part 1 (Warsaw, 1988) (Warsaw), Banach Center Publ., no. 26, PWN, 1990, With an appendix by William Crawley-Boevey, pp. 407–432. MR 93e:16022

93. _____, *Recent advances in the representation theory of finite-dimensional algebras*, Representation theory of finite groups and finite-dimensional algebras (Bielefeld, 1991), Progr. Math., vol. 95, Birkhäuser, Basel, 1991, pp. 141–192. MR 92i:16012

94. _____, *The braid group action on the set of exceptional sequences of a hereditary Artin algebra*, Abelian group theory and related topics (Oberwolfach, 1993), Contemp. Math., vol. 171, Amer. Math. Soc., Providence, RI, 1994, pp. 339–352. MR 95m:16006

95. _____, *Continued fractions, tilting modules and the construction of large indecomposable modules*, 2002, Talk at the ICRA X, Toronto, July 2002.

96. L. Rowen, *Ring theory (vol. 1 and 2)*, Pure and Applied Mathematics, vol. 127/128, Academic Press Inc., Boston, MA, 1988. MR 89h:16001/16002

97. A. N. Rudakov, *Exceptional vector bundles on a quadric*, Izv. Akad. Nauk SSSR Ser. Mat. **52** (1988), no. 4, 788–812, 896. MR 90a:14019

98. _____, *Markov numbers and exceptional bundles on* $\mathbf{P}^2$, Izv. Akad. Nauk SSSR Ser. Mat. **52** (1988), no. 1, 100–112, 240. MR 89f:14012

99. P. Samuel, *Lectures on unique factorization domains*, Notes by M. Pavman Murthy. Tata Institute of Fundamental Research Lectures on Mathematics, No. 30, Tata Institute of Fundamental Research, Bombay, 1964. MR 35 #5428

100. I. N. Sanov, *A property of a representation of a free group*, Doklady Akad. Nauk SSSR (N. S.) **57** (1947), 657–659. MR 9,224e

101. P. Seidel and R. Thomas, *Braid group actions on derived categories of coherent sheaves*, Duke Math. J. **108** (2001), no. 1, 37–108. MR 2002e:14030

102. J.-P. Serre, *Faisceaux algébriques cohérents*, Ann. of Math. (2) **61** (1955), 197–278. MR 16,953c

103. C. S. Seshadri, *Fibrés vectoriels sur les courbes algébriques*, Astérisque, vol. 96, Société Mathématique de France, Paris, 1982, Notes written by J.-M. Drezet from a course at the École Normale Supérieure, June 1980. MR 85b:14023

104. A. Skowroński, *On omnipresent tubular families of modules*, Representation Theory of Algebras (Cocoyoc, 1994) (R. Bautista, R. Martínez-Villa, and J. A. de la Peña, eds.), CMS Conf. Proc., vol. 18, Amer. Math. Soc., Providence, RI, 1996, pp. 641–657. MR 97f:16032

105. J. T. Stafford and M. van den Bergh, *Noncommutative curves and noncommutative surfaces*, Bull. Amer. Math. Soc. (N.S.) **38** (2001), no. 2, 171–216 (electronic). MR 2002d:16036

106. J.-P. van Deuren, J. Van Geel, and F. Van Oystaeyen, *Genus and a Riemann-Roch theorem for noncommutative function fields in one variable*, Paul Dubreil and Marie-Paule Malliavin Algebra Seminar, 33rd Year (Paris, 1980), Lecture Notes in Math., vol. 867, Springer, Berlin, 1981, pp. 295–318. MR 83k:14007

107. O. Venjakob, *A non-commutative Weierstrass preparation theorem and applications to Iwasawa theory*, J. Reine Angew. Math. **559** (2003), 153–191, With an appendix by Denis Vogel. MR 2004e:11123

Index

Editorial Information

To be published in the *Memoirs*, a paper must be correct, new, nontrivial, and significant. Further, it must be well written and of interest to a substantial number of mathematicians. Piecemeal results, such as an inconclusive step toward an unproved major theorem or a minor variation on a known result, are in general not acceptable for publication.

Papers appearing in *Memoirs* are generally at least 80 and not more than 200 published pages in length. Papers less than 80 or more than 200 published pages require the approval of the Managing Editor of the Transactions/Memoirs Editorial Board.

As of May 31, 2009, the backlog for this journal was approximately 11 volumes. This estimate is the result of dividing the number of manuscripts for this journal in the Providence office that have not yet gone to the printer on the above date by the average number of monographs per volume over the previous twelve months, reduced by the number of volumes published in four months (the time necessary for preparing a volume for the printer). (There are 6 volumes per year, each usually containing at least 4 numbers.)

A Consent to Publish and Copyright Agreement is required before a paper will be published in the *Memoirs*. After a paper is accepted for publication, the Providence office will send a Consent to Publish and Copyright Agreement to all authors of the paper. By submitting a paper to the *Memoirs*, authors certify that the results have not been submitted to nor are they under consideration for publication by another journal, conference proceedings, or similar publication.

Information for Authors

Memoirs are printed from camera copy fully prepared by the author. This means that the finished book will look exactly like the copy submitted.

Initial submission. The AMS uses Centralized Manuscript Processing for initial submissions. Authors should submit a PDF file using the Initial Manuscript Submission form found at `www.ams.org/peer-review-submission`, or send one copy of the manuscript to the following address: Centralized Manuscript Processing, MEMOIRS OF THE AMS, 201 Charles Street, Providence, RI 02904-2294 USA. If a paper copy is being forwarded to the AMS, indicate that it is for it Memoirs and include the name of the corresponding author, contact information such as email address or mailing address, and the name of an appropriate Editor to review the paper (see the list of Editors below).

The paper must contain a *descriptive title* and an *abstract* that summarizes the article in language suitable for workers in the general field (algebra, analysis, etc.). The *descriptive title* should be short, but informative; useless or vague phrases such as "some remarks about" or "concerning" should be avoided. The *abstract* should be at least one complete sentence, and at most 300 words. Included with the footnotes to the paper should be the 2000 *Mathematics Subject Classification* representing the primary and secondary subjects of the article. The classifications are accessible from `www.ams.org/msc/`. The list of classifications is also available in print starting with the 1999 annual index of *Mathematical Reviews*. The Mathematics Subject Classification footnote may be followed by a list of *key words and phrases* describing the subject matter of the article and taken from it. Journal abbreviations used in bibliographies are listed in the latest *Mathematical Reviews* annual index. The series abbreviations are also accessible from `www.ams.org/msnhtml/serials.pdf`. To help in preparing and verifying references, the AMS offers MR Lookup, a Reference Tool for Linking, at `www.ams.org/mrlookup/`.

Electronically prepared manuscripts. The AMS encourages electronically prepared manuscripts, with a strong preference for $\mathcal{A}\mathcal{M}\mathcal{S}$-LaTeX. To this end, the Society has prepared $\mathcal{A}\mathcal{M}\mathcal{S}$-LaTeX author packages for each AMS publication. Author packages include instructions for preparing electronic manuscripts, samples, and a style file that generates the particular design specifications of that publication series. Though $\mathcal{AMS}$-LATEX is the highly preferred format of TEX, author packages are also available in $\mathcal{AMS}$-TEX.

Authors may retrieve an author package for *Memoirs of the AMS* from www.ams.org/journals/memo/memoauthorpac.html or via FTP to ftp.ams.org (login as anonymous, enter username as password, and type cd pub/author-info). The *AMS Author Handbook* and the *Instruction Manual* are available in PDF format from the author package link. The author package can also be obtained free of charge by sending email to tech-support@ams.org (Internet) or from the Publication Division, American Mathematical Society, 201 Charles St., Providence, RI 02904-2294, USA. When requesting an author package, please specify $\mathcal{AMS}$-LATEX or $\mathcal{AMS}$-TEX and the publication in which your paper will appear. Please be sure to include your complete mailing address.

After acceptance. The final version of the electronic file should be sent to the Providence office (this includes any TEX source file, any graphics files, and the DVI or PostScript file) immediately after the paper has been accepted for publication.

Before sending the source file, be sure you have proofread your paper carefully. The files you send must be the EXACT files used to generate the proof copy that was accepted for publication. For all publications, authors are required to send a printed copy of their paper, which exactly matches the copy approved for publication, along with any graphics that will appear in the paper.

Accepted electronically prepared files can be submitted via the web at www.ams.org/submit-book-journal/, sent via FTP, or sent on CD-Rom or diskette to the Electronic Prepress Department, American Mathematical Society, 201 Charles Street, Providence, RI 02904-2294 USA. TEX source files, DVI files, and PostScript files can be transferred over the Internet by FTP to the Internet node ftp.ams.org (130.44.1.100). When sending a manuscript electronically via CD-Rom or diskette, please be sure to include a message identifying the paper as a Memoir.

Electronically prepared manuscripts can also be sent via email to pub-submit@ams.org (Internet). In order to send files via email, they must be encoded properly. (DVI files are binary and PostScript files tend to be very large.)

Electronic graphics. Comprehensive instructions on preparing graphics are available at www.ams.org/authors/journals.html. A few of the major requirements are given here.

Submit files for graphics as EPS (Encapsulated PostScript) files. This includes graphics originated via a graphics application as well as scanned photographs or other computer-generated images. If this is not possible, TIFF files are acceptable as long as they can be opened in Adobe Photoshop or Illustrator. No matter what method was used to produce the graphic, it is necessary to provide a paper copy to the AMS.

Authors using graphics packages for the creation of electronic art should also avoid the use of any lines thinner than 0.5 points in width. Many graphics packages allow the user to specify a "hairline" for a very thin line. Hairlines often look acceptable when proofed on a typical laser printer. However, when produced on a high-resolution laser imagesetter, hairlines become nearly invisible and will be lost entirely in the final printing process.

Screens should be set to values between 15% and 85%. Screens which fall outside of this range are too light or too dark to print correctly. Variations of screens within a graphic should be no less than 10%.

Inquiries. Any inquiries concerning a paper that has been accepted for publication should be sent to memo-query@ams.org or directly to the Electronic Prepress Department, American Mathematical Society, 201 Charles St., Providence, RI 02904-2294 USA.

Titles in This Series

TITLES IN THIS SERIES

For a complete list of titles in this series, visit the
AMS Bookstore at **www.ams.org/bookstore/**.